Evaluation of Ozone and Ultraviolet Light

Awwa Research Foundation

About the Awwa Research Foundation

The Awwa Research Foundation (AwwaRF) is a member-supported, international, nonprofit organization that sponsors research to enable water utilities, public health agencies, and other professionals to provide safe and affordable drinking water to consumers.

The Foundation's mission is to advance the science of water to improve the quality of life. To achieve this mission, the Foundation sponsors studies on all aspects of drinking water, including supply and resources, treatment, monitoring and analysis, distribution, management, and health effects. Funding for research is provided primarily by subscription payments from approximately 1,000 utilities, consulting firms, and manufacturers in North America and abroad. Additional funding comes from collaborative partnerships with other national and international organizations, allowing for resources to be leveraged, expertise to be shared, and broad-based knowledge to be developed and disseminated. Government funding serves as a third source of research dollars.

From its headquarters in Denver, Colorado, the Foundation's staff directs and supports the efforts of more than 800 volunteers who serve on the board of trustees and various committees. These volunteers represent many facets of the water industry, and contribute their expertise to select and monitor research studies that benefit the entire drinking water community.

The results of research are disseminated through a number of channels, including reports, the Web site, conferences, and periodicals.

For subscribers, the Foundation serves as a cooperative program in which water suppliers unite to pool their resources. By applying Foundation research findings, these water suppliers can save substantial costs and stay on the leading edge of drinking water science and technology. Since its inception, AwwaRF has supplied the water community with more than $300 million in applied research.

More information about the Foundation and how to become a subscriber is available on the Web at **www.awwarf.org**.

Evaluation of Ozone and Ultraviolet Light

Prepared by:
Robert A. Hulsey, **Albert J. Capuzzi**, **Catherine M. Spencer**
Black & Veatch, Kansas City, MO 64114
and
James P. Malley and **Steven Farabaugh**
University of New Hampshire, Durham, NH

Jointly sponsored by:
Awwa Research Foundation
6666 West Quincy Avenue, Denver, CO 80235-3098

Massachusetts Water Resources Authority
Charlestown Navy Yard, Boston, MA 02129

Concord Public Works
Concord, MA 01742
and

Southern Nevada Water Authority
Las Vegas, NV 89153

Published by:

Awwa Research Foundation
Advancing the Science of Water®

DISCLAIMER

This study was jointly funded by the Awwa Research Foundation (AwwaRF), the Massachusetts Water Resources Authority (MWRA), the Concord Public Works (Concord), and the Southern Nevada Water Authority (SNWA). AwwaRF, MWRA, Concord, and SNWA assume no responsibility for the content of the research study reported in this publication or for the opinions or statements of fact expressed in the report. The mention of trade names for commercial products does not represent or imply the approval or endorsement of either AwwaRF, MWRA, Concord, or SNWA. This report is presented solely for informational purposes.

CONTENTS

TABLES

FIGURES

FOREWORD

The Awwa Research Foundation is a nonprofit corporation that is dedicated to the implementation of a research effort to help utilities respond to regulatory requirements and traditional high-priority concerns of the industry. The research agenda is developed through a process of consultation with subscribers and drinking water professionals. Under the umbrella of a Strategic Research Plan, the Research Advisory Council prioritizes the suggested projects based upon current and future needs, applicability, and past work; the recommendations are forwarded to the Board of Trustees for final selection. The foundation also sponsors research projects through the unsolicited proposal process; the Collaborative Research, Research Applications, and Tailored Collaboration programs; and various joint research efforts with organizations such as the U.S. Environmental Protection Agency, the U.S. Bureau of Reclamation, and the Association of California Water Agencies.

This publication is a result of one of these sponsored studies, and it is hoped that its findings will be applied in communities throughout the world. The following report serves not only as a means of communicating the results of the water industry's centralized research program but also as a tool to enlist the further support of the nonmember utilities and individuals.

Projects are managed closely from their inception to the final report by the foundation's staff and large cadre of volunteers who willingly contribute their time and expertise. The foundation serves a planning and management function and awards contracts to other institutions such as water utilities, universities, and engineering firms. The funding for this research effort comes primarily from the Subscription Program, through which water utilities subscribe to the research program and make annual payment proportionate to the volume of water they deliver, and consultants and manufacturers subscribe based on their annual billings. The program offers a cost-effective and fair method for funding research in the public interest.

A broad spectrum of water supply issues is addressed by the foundation's research agenda: resources, treatment and operations, distribution and storage, water quality and analysis, toxicology, economics, and management. The ultimate purpose of the coordinated effort is to assist water suppliers to provide the highest possible quality of water economically and reliably. The true benefits are realized when the results are implemented at the utility level. The foundation's trustees are pleased to offer this publication as a contribution toward that end.

Walter J. Bishop
Chair, Board of Trustees
Awwa Research Foundation

James F. Manwaring, P.E.
Executive Director
Awwa Research Foundation

ACKNOWLEDGMENTS

The authors of this report are indebted to the following water utilities, university participants, and individuals for their cooperation and participation in this project:

Massachusetts Water Resources Authority, Capital Engineering and Construction Department, Jae Kim and Tiffany Tran
Western Operations Department, James Muri, construction and operation team
Laboratory, Windsor Sung
Concord Public Works, Water /Sewer Division, Alan Cathcart, Tom Ford, Matt Mostoller and Elena Proakis
Southern Nevada Water Authority, David Rexing and Eric Wert
Center for Biofilm Engineering, Montana State University, Bozeman, MT, Anne Camper and Kelli Buckingham-Meyer
University of New Hampshire, David Gaithuma
University of Arizona, Marilyn Marshall
Clancy Environmental Consultants, Jennifer Clancy, Tom Harrgy
Tufts University School of Veterinary Medicine, Dr. Widmer
Mitsubishi Electric, Paul Rutherford
University of Massachusetts, Amherst, David Reckhow and Narayan Venkatesan
Trojan Technologies, Michael Shortt
Portland Water Bureau

We would also like to acknowledge the help and advice of the Project Advisory Committee (PAC) comprised of Manouchehr Boozapour, Robert Cushing, William Soucie and AwwaRF project officer Bob Allen.

The authors wish to thank Anthony Accardi, Dominick Loschiavo and David Goncalves' for managing procurement and construction of pilot facilities, Gerald Benson for supporting the project, Jessica Edwards-Brandt and Heather Mackey for technical assistance Alicia Moyer for typing and Jonathan Clement whose idea initiated this effort.

EXECUTIVE SUMMARY

The primary goal of this research was to evaluate ozone application followed by ultraviolet light (UV) to determine how to best operate the two disinfection systems efficiently for protection of the public health. Protozoan parasites have become a major public health concern in drinking water systems due to waterborne outbreaks in the late 1980's and early 1990's. The requirement for two log Cryptosporidium inactivation outlined in the Interim Enhanced Surface Water Treatment Rule (ESWTR) was expanded in the proposed Long Term 2 ESWTR so that utilities were required to develop means to inactivate *Cryptosporidium* in correspondence with the levels of these organisms in the specific source water. Unfiltered surface water systems, in particular, will need to combine the strengths of different types of disinfectants to meet the requirements for *Cryptosporidium* inactivation.

RESEARCH OBJECTIVES

Stakeholders with particular interest in UV technologies include large water utilities which have already installed ozone for *Giardia* or *Cryptosporidium* inactivation, taste and odor control, and/or oxidation, and are interested in increased pathogen protection at lower operating costs; and unfiltered water systems with sole reliance on disinfection for pathogen inactivation. However, several important questions about UV for these and other water utilities remained, including:

1) Can medium pressure UV light (MP-UV) alone or the combination of ozone and medium pressure UV light (MP-UV) cost effectively inactivate *Giardia*, *Cryptosporidium*, and viruses, especially under cold water conditions?
2) Will MP-UV affect disinfection by-product formation or DBP precursors?
3) What effect will MP-UV have on the biostability/regrowth potential of the resulting water in the distribution system?

To answer these questions, a comprehensive study protocol was developed. The tasks under the protocol were:

1. A literature review was completed to analyze available information that pertains to ultraviolet light disinfection and ozone inactivation of protozoa, bacteria, and viruses as well as current design guidelines for these disinfection processes.
2. Bench scale studies were completed to verify the level of inactivation accomplished by ozone against *Cryptosporidium* in cold water; to expand knowledge regarding inactivation of *Giardia* and *Cryptosporidium* using ultraviolet light in cold water, and to do bench-scale collimated beam studies using MS-2 bacteriophage to validate the UV dose being provided in the pilot scale biodosimetry studies conducted in the field.
3. Field studies were conducted to demonstrate performance of ultraviolet light disinfection and ozone followed by ultraviolet light disinfection on MS-2 surrogates under different water temperature and UV lamp age conditions. The field studies also looked disinfection by-product formation, bacterial regrowth, and ultraviolet absorbance of the treated water.
4. Development of the operational strategies to determine an optimal balance of ozone dose and UV for disinfection. The strategies were based on the bench scale and field data, as well as operating conditions coupled with capital and operating cost information.

FINDINGS

The bench scale *Cryptosporidium* work confirmed prior studies indicating that ozone alone is not effective for inactivation of *Cryptosporidium* at low water temperatures. The bench scale *Giardia* work demonstrated that ultraviolet light disinfection was very effective for inactivation of *Giardia* at low water temperatures. The tests were conducted at 1°C. The application of ozone followed by ultraviolet light resulted in the inactivation of *Cryptosporidium* at greater than 99.999% (5 log) and *Giardia* at greater than 99.9% (3 log).

The field demonstration of ozone followed by UV irradiation for inactivation of the MS2 surrogate revealed greater than 5 log inactivation with new lamps in 8°C water. A second challenge trial was conducted sixteen months later, at the end of normal lamp life. The second challenge trial revealed 2 log inactivation of the MS2 surrogate with the aged lamps.

There were significant differences between the pre-ozonated treatment train compared to the non-ozonated treatment train with respect to trihalomethane and haloacetic acid concentrations with the pre-ozonated samples at the influent and effluent of the pipe loops consistently lower than their non-ozonated counterparts. Overall, pre-ozonation conferred lower levels of halogenated disinfection by-products with this water source compared to the non-ozonated water. Keto-acids were the only disinfection by-product tested where the levels in the ozonated water were consistently higher than in the non-ozonated water. No significant differences were observed between the attached heterotrophic plate count microbial numbers for pre-ozonated, UV irradiated, post treated water compared to non-ozonated, UV irradiated, post treated water.

This research study has found that combining the strengths of ozone for oxidation, taste and odor reduction and virus inactivation, and UV for inactivation of protozoa can provide a cost-effective treatment strategy. Placing the UV after ozone addition resulted in UV capital costs savings of 19 to 21 percent compared to UV as a stand alone technology for MWRA and similar large, unfiltered surface water systems. The improvement in UVT afforded by ozone application results in UV units that require fewer lamps to achieve the target inactivation of *Cryptosporidium* required. Operating costs for the large unfiltered surface water ultraviolet system were 29 to 32 percent lower post ozone than with no ozone or ahead of ozone.

CHAPTER 1
STUDY FINDINGS AND RECOMMENDATIONS

INTRODUCTION

The primary goal of this research was to evaluate the benefits of Ultraviolet light (UV) and ozone disinfection and to determine how to best operate the two disinfection systems efficiently for protection of public health. Protozoan parasites have become a major public health concern in drinking water systems due to waterborne outbreaks in the late 1980's and early 1990's. US Environmental Protection Agency (USEPA) regulators responded to these outbreaks with new regulations such as the Surface Water Treatment Rule (SWTR), the Interim Enhanced Surface Water Treatment Rule (IESWTR), the Long Term 1 Enhanced Surface Water treatment Rule (LT1ESWTR) and the Long Term 2 Surface Water Treatment Rule (LT2ESWTR) which is currently in the proposed stage. The requirement for two log *Cryptosporidium* inactivation outlined in the Interim ESWTR was expanded in the proposed LT2ESWTR so that utilities were required to develop site specific means to remove or inactivate *Cryptosporidium* corresponding to the levels of these organisms in the source water. Filtration may be expected to remove 2.5 to 3 log concentrations of protozoa; thus, the LT2ESWTR may require further removal or inactivation and provides guidance in the form of a toolbox (USEPA, 2003). Table 1.1 presents the bin requirements for filtered plants.

Table 1.1
Bin requirements for filtered plants

Cryptosporidium cysts/liter	Bin classification	Conventional filtration	Direct filtration	Slow sand or DE filtration	Total Inactivation
<0.075	1	No additional treatment required			
≥ 0.075 and <1.0	2	1 log	1.5 log	1 log	4.0 log[1]
≥1.0 and <3.0	3	2 log[2]	2.5 log [2]	2 log[2]	5.0 log[1]
≥ 3.0	4	2.5 log[2]	3 log[2]	2.5 log [2]	5.5 log[1]

Source: adapted from USEPA, 2003.

[1] As determined by the State

[2] Systems must achieve 1 log treatment using technologies listed below

USEPA requires that at least 1 log treatment must be provided using

- Ozone
- Chlorine dioxide
- Ultraviolet light
- Membranes
- Bag or cartridge filters
- Bank Filtration

for those systems that must provide 4 log inactivation or more. Of these technologies, ultraviolet light disinfection has elicited the most interest as a relatively novel option for treatment of surface waters. For many, the promise of UV is that of a low-cost technology, which would allow utilities to comply with regulatory requirements that specifically call for *Cryptosporidium* inactivation at 2 or more logs.

Research Objectives

Stakeholders with particular interest in UV technologies include large water utilities which have already installed ozone for *Giardia* or *Cryptosporidium* inactivation, taste and odor control, and/or oxidation, and are interested in increased pathogen protection at lower operating costs; and unfiltered water systems with sole reliance on disinfection for pathogen inactivation. However, several important questions about UV for these and other water utilities remained, including:

1) Can medium pressure UV light (MP-UV) or the combination of ozone and medium pressure UV light (MP-UV) cost effectively meet concentration over time (CT) requirements for *Giardia*, *Cryptosporidium*, and viruses, especially under cold water conditions?

2) What effects will the combination of ozone followed by MP-UV have on disinfection by-product formation or DBP precursors? Will MP-UV affect disinfection by-product formation or DBP precursors?

3) What effect will MP-UV have on the biostability/regrowth potential of the resulting water in the distribution system? What effects will the combination of ozone followed by MP-UV have on the biostability/ regrowth potential of the resulting water in the distribution system?

To answer these questions, the following tasks were completed:

- A literature review was completed in order to analyze available information that pertains to ultraviolet light disinfection and ozone inactivation of protozoa, bacteria, and viruses as well as current design guidelines for these disinfection processes. This background information was used as a starting point for the bench scale and field investigations.

- Bench scale studies were completed to verify the level of inactivation accomplished by ozone against *Cryptosporidium* in cold water; to expand knowledge regarding inactivation of *Giardia* using ultraviolet light in cold water, and to demonstrate inactivation ability of ozone followed by ultraviolet light for both *Cryptosporidium* and *Giardia*. Also, it was necessary to do bench-scale collimated beam studies to validate the UV dose being provided in the pilot scale biodosimetry studies conducted in the field. Therefore, bench scale studies were conducted using batch ozonation and collimated beam UV equipment with MS2 bacteriophage.

- Field studies were conducted to demonstrate performance of ultraviolet light disinfection and ozone followed by ultraviolet light disinfection on MS-2 surrogates under different water temperature and UV lamp age conditions. The field studies also looked at other water quality issues such as disinfection by-product formation, bacterial regrowth, ultraviolet absorbance of the treated water, and potential effects on corrosion control.

- Review of the bench scale and field data, operating conditions coupled with capital and operating cost information was undertaken to determine the appropriate treatment load-sharing among the two primary disinfectant technologies. Specifically, the most cost-effective combination of ozone and UV to share pathogen inactivation requirements and disinfection by-product reduction is presented. Recommendations for integration of UV into systems with existing ozone applications were developed.

FINDINGS FROM BENCH SCALE WORK

The *Cryptosporidium* inactivation trials completed by Clancy Environmental confirmed prior studies indicating that ozone alone is not effective for inactivation of *Cryptosporidium* at low water temperatures. The tests were conducted at $1^{o}C$. Two ozone doses, 1.5 milligram per liter (mg/L) and 2.5 mg/L, were applied to the cold water for a contact time of 40 minutes to provide CT values of 24 mg*min/L and 37 mg*min/L respectively. Both of the ozone applications resulted in *Cryptosporidium* inactivation of less than 90 percent (1 log).

The *Giardia* inactivation work completed by the University of New Hampshire and Tufts University demonstrated that UV disinfection was very effective for inactivation of *Giardia* at low water temperatures. The tests were also conducted at $1^{o}C$. Two UV doses of 5 milliJoule per square centimeter (mJ/cm^2) and 10 mJ/cm^2 were applied to the water, resulting in greater than 99.9 percent (3 log) inactivation of the *Giardia* for each dose.

The application of ozone followed by ultraviolet light resulted in the inactivation of *Cryptosporidium* at greater than 99.999% (5 log) and *Giardia* at greater than 99.9% (3 log). Ozone doses and contact times at 1.5 mg/L and 40 minutes were the same for both microorganisms. The UV dose applied to the *Cryptosporidium* was 5 mJ/cm^2, while the UV dose applied to the *Giardia* was 10 mJ/cm^2.

FINDINGS FROM THE FIELD WORK

Two sites were used for field trials of ozone followed by UV irradiation for primary disinfection. Massachusetts Water Resources Authority (MWRA) constructed a 400 gallon per minute pilot scale facility to house their model distribution system piping (Clement et al.,2002). Half of the pipe loops received water that was treated with medium pressure UV followed by chlorine, pH and alkalinity adjustment and ammonia to form chloramines. The other half of the pipe loops received water that was ozonated, treated with medium pressure UV followed by chlorine, pH and alkalinity adjustment and ammonia to form chloramines. A comparison between the treated water entering and exiting the pipe loops was made to find differences that could be attributed to ozone/UV versus UV alone.

The Town of Concord, Massachusetts retrofitted their existing Nagog Pond Ozone Facility with a medium pressure ultraviolet unit, treating all of the surface water that entered the distribution system from that source. Operating data from the Ozone Facility and water quality data from the distribution system were evaluated to determine the effect of UV irradiation on water quality.

Transmittance

At the MWRA field site, no significant differences were seen between the raw, pre-ozonated, UV irradiated, and non-ozonated, UV irradiated water with regard to total organic carbon concentrations. However, significant differences were seen in the ultraviolet absorbance of the pre-ozonated water compared to the raw water and the non-ozonated water. Regardless of the season, the addition of ozone improved the UV transmittance of the water, particularly in the critical range of 220 to 300 nanometers. The Concord field data also demonstrated significant improvement in ultraviolet transmittance of the water after ozone application.

Challenge Trials

Two challenge trials were conducted at MWRA during the course of the study. The first trial, conducted in October 2002 demonstrated MS2 inactivation using new lamps for both the pre-ozonated and non –ozonated trains. However, there was some ozone carryover from the pre-ozonated train and MS-2 injected after ozone contact were destroyed by the ozone before they were irradiated by the UV. Data were obtained from the non-ozonated train. The second trial, conducted in June 2003, demonstrated MS-2 inactivation of the aged lamps. The ozone system was inoperable on the date of the second trial so no useful data could be obtained comparing pre-ozonated and non-ozonated water at this site. Table 1.2 presents a summary of the challenge trial findings for the non-ozonated train of the MWRA pilot.

Table 1.2
Summary of MWRA challenge trial data
Non-ozonated train

Lamp power %	October MS-2 log reduction	June MS-2 log reduction	October RED[1] mJ/cm^2	June RED mJ/cm^2
100	3.25	1.92	72.2	46
67.3	2.40	1.10	51.9	26.4
30	1.06	0.74	19.9	17.4

[1] RED is reduction equivalent dose

Both of the challenge trials conducted at the Concord site looked at pre-ozonated and non-ozonated water, thus comparisons could be made between the pre-ozonated and non-ozonated inactivation rates (Table 1.3). Ozone doses were similar for both trials. During the first trial in March, 2002, a small but not statistically significant difference between the preozonated and non ozonated samples was observed, suggesting slightly better reduction equivalent dose at 30 percent lamp power after ozonation, however, inactivation rates for both modes of ozone/UV operation at the higher lamps powers was so great that differences could not be detected. Again, during the second challenge trial conducted in June 2003, a small but not significantly different difference between the preozonated and non ozonated samples was observed at all lamp power levels. For the second trial, the lamps and sensors were quite aged and with only three data points, the low power setting reading skewed the results. At the low power setting, nearly all of the UV radiation was attenuated and little inactivation was observed. The increase in transmittance with ozone didn't offset the low wattage of the lamp and the distance from the lamp to the sensor.

Table 1.3
Summary of Concord challenge trial data

	Lamp Power %	March MS-2 log reduction	June MS-2 log reduction	March RED mJ/cm^2	June RED mJ/cm^2
Ozone dose, mg/L		2.96	2.84		
Ozonated	100	>5.89	2.30	>110	43
	67.3	>5.90	1.27	>110	21.5
	30	4.39	0.64	73.4	8.4
Non-ozonated	100	>5.86	2.12	>110	39
	67.3	>5.86	1.00	>110	15.7
	30	4.29	0.62	70.9	8.4

Disinfection By-Products

There were significant differences between the pre-ozonated treatment train compared to the non-ozonated treatment train with respect to trihalomethane (THM) and haloacetic acid (HAA) formation. Pre-oxidation with ozone resulted in lower THM and HAA concentrations than UV-only treatment for samples from both the inlet and outlet of the model distribution system While a change was noted in THM and HAA formation, the concentration of total organic halides (TOX) was the same at the entrance to the model distribution piping regardless of primary disinfection scheme. However, the pre-ozonated treatment train developed lower TOX levels within the model distribution system compared to the non-ozonated treatment train. Overall, pre-ozonation resulted in lower levels of halogenated disinfection by-products with this water source compared to the non-ozonated water. Bromide levels are low for both the Wachusett and Nagog Pond sources so bromate data were not obtained.

Ozonation clearly increased the levels of aldehydes and keto-acids. However, the aldehyde levels rapidly degraded in the pre-ozonated treatment train such that after chloramination, both the pre-ozonated and non-ozonated aldehyde concentrations were similar. Keto-acids also rapidly degraded though the levels remained higher in the pre-ozonated samples after chloramination and passage through the pipe loops than in the non-ozonated samples.

Regrowth

No significant differences were observed between the attached heterotrophic plate count microbial numbers for pre-ozonated, UV irradiated, post treated water compared to non-ozonated, UV irradiated, post treated water. While there were some differences in the bulk water microbial numbers between the two treatments entering their respective pipe loops, the effluent bulk water heterotrophic plate counts were not significantly different.

DESIGN AND OPERATIONAL GUIDANCE

This research study has found that combining the strengths of ozone (oxidation, taste and odor reduction and virus inactivation) and UV (inactivation of *Cryptosporidium*) can provide a cost-effective treatment strategy. Ozone and UV have advantages and disadvantages, which are summarized in Table 1.4. However, using both disinfectants in combination at a water treatment plant allows each system to be optimized, utilizing the benefits of both technologies. Ozone and UV together can satisfy a water work's oxidation and inactivation requirements while producing very low formation of regulated disinfection byproducts and could also result in savings in operation costs, especially energy.

Table 1.4
Benefits and detractions of ozone and UV

	Ozone	UV
Bacteria and viruses	Excellent	Good
Protozoa	Good	Excellent
Regulated DBP formation*	Low	Few at required dose
Residual	None	None
Oxidant	Very good	Only with H_2O_2
Microflocculant	Yes	No effect
Capital cost	High	Low
Operating cost	Low	Low

* bromate formation may be a problem for some waters

One of the benefits of utilizing both ozone and UV is that they would provide an excellent multiple-barrier treatment approach. Ozone followed by UV provides a redundant system for *Giardia* inactivation and potentially for *Cryptosporidium* inactivation (under certain conditions). Also, if the ozone system were to go offline, the UV system could meet the inactivation requirements effectively albeit by bringing on a spare UV reactor or reducing flow rate through the plant. A strong oxidant like ozone can also increase the UV transmittance of the water, which can result in a smaller, more affordable UV reactor.

Placing UV after ozone addition resulted in UV capital costs savings of approximately 20 percent compared to UV as a stand alone technology for MWRA (~220 mgd) and similar large, unfiltered surface water systems. The improvement in UVT afforded by ozone application results in UV reactors that require fewer lamps to achieve the target inactivation of *Cryptosporidium* required. Fewer lamps resulted in lower energy and maintenance costs and less labor for post ozone UV reactors. Operating costs for the large unfiltered surface water UV system were approximately 30 percent lower post ozone than with no ozone or ahead of ozone.

The size of the Concord facility (~ 1 mgd) and water demands are such that no substantial differences were proposed by UV manufacturers for units treating raw water versus ozonated water even though there is a significant improvement in UVT of the ozonated water.

FINDINGS

The research found that:

1. Medium pressure UV light disinfection, either with or without ozone, is capable of cost effectively meeting CT or, more properly, IT requirements for the inactivation of *Giardia* and *Cryptosporidium* under cold water conditions.

2. The use of ozone resulted in lower levels of halogenated disinfection by-products. For the unfiltered water that was tested, the combination of ozone followed by UV irradiation resulted in lower levels of halogenated disinfection by-products after chlorination compared to water treated with UV irradiation as the primary disinfectant and chlorine as the secondary disinfectant. MP-UV did not appear to affect disinfection by-product precursors.

3. For the unfiltered water that was tested, there were no differences in the regrowth potential between water subjected to ozone followed by MP-UV and water subjected to MP-UV alone as the primary disinfectant.

Using ozonation followed by UV in a combined disinfection scenario allows both disinfectants to work in a complimentary way. Some of these benefits are listed below:

- Ozone is a chemical disinfectant, UV is a physical disinfectant. They act in two very different ways to inactivate microorganisms and together create formidable barriers to the passage of viable protozoa, bacteria and viruses into the finished water.

- In cold water, very high ozone doses and CT values are required to disinfect *Cryptosporidium* cysts. Ozone systems would require extremely long detention times and high ozone residuals concentrations to provide the necessary CT at water temperatures below 5 °C. UV disinfection on the other hand, is insensitive to water temperature.

- Ozone is effective at disinfecting viruses while much higher UV doses would be required based on Adenovirus data in the Ultraviolet Disinfection Guidance Manual (UVDGM).

- Ozone can provide collateral water quality benefits such as taste and odor elimination, oxidation of dissolved iron and manganese, and destruction of disinfection by-product precursors.

- Ozonation increases a water's UV transmittance, thereby making UV disinfection more effective.

- At high doses, ozone can increase the formation of biodegradable organic matter such as aldehydes and keto-acids, or bromate. By using lower ozone doses in a combined Ozone/UV disinfection scheme, formation of these by-products can be reduced.

- From a design perspective, looking strictly at the capital cost of the UV equipment, a post-ozone UV facility is estimated to be about 20% less expensive than a pre-ozone or no ozone UV facility. This benefit accrues primarily to large systems where substantial savings in numbers of UV units or lamps is seen as the transmittance of the water improves.

- Improved operational controls for UV units may permit utilities to operate ozone and UV more beneficially. The pilot UV units were limited to set increments of lamp power, often providing more UV light and using more electrical power than was necessary to achieve 2 to 3 log *Cryptosporidium* inactivation. More operational control of UV doses with a minimum lamp power at a setting that will ensure the minimum required inactivation would be a tool to save operational costs.

There are some potential disadvantages of Post-Ozone UV Systems:

- When UV reactors are placed after ozonation, care must be taken to ensure that ozone residual is minimized prior to UV. Excessive ozone residual will be a source of ultraviolet absorbance and lower the disinfection effectiveness of the UV.

- If, because of power fluctuations, the UV lamps should go off, the ozone dosage could be increased to inactivate *Cryptosporidium*. However, there will be some water that passes to the distribution system before this adjustment occurs. Thus, off-specification water will pass into the clearwell.

- Fouling by inorganic iron and manganese, if present in the raw water, may be increased by oxidation of both of these metals by ozone ahead of UV disinfection.

RECOMMENDATIONS

Systems that already have ozone can benefit greatly by including the increase in transmittance into the design of the UV system, making it smaller and more energy efficient. As will all pretreatment systems, the goal of operations would be to maintain ozonation ahead of the UV system. A utility considering UV following an existing ozone system should evaluate the ozone system down time during the year. Is the amount of time acceptable and could it be counted toward the off-specification allotment?

In general, systems that don't have ozone and are considering UV would not find the savings in UV system costs to be enough to pay for the ozone system. If ozone is used for other purposes besides increasing transmittance (a second line of defense in a multi-barrier approach to pathogen control, taste and odor control, oxidation, filtration enhancement, color removal, THM/HAA reduction), the utility can see a benefit by installing both technologies. If a new ozone system were to be installed with a UV system, then ozone system redundancy should be evaluated.

For both existing and new ozone systems, the operations strategy needs to be defined. Two potential strategies include using the ozone system to provide some level of disinfection and UV as another disinfection barrier or using the ozone more as pretreatment for the UV which becomes the primary disinfectant. If the ozone were to go down, the down may be included in the off-specification time allotted. The UV reactor design should accommodate the range of flows expected and define acceptable operating ranges with ozone on-line and off.

For smaller systems, the improvement in UVT with ozone application may not translate to lower capital costs as there are limits to design flexibility for both systems. However, the operational and regulatory compliance benefits of ozone and UV outlined above can be seen by small systems.

CHAPTER 2
LITERATURE REVIEW OF TREATMENT TECHNOLOGIES FOR ALTERNATIVE DISINFECTANTS

Disinfection is the inactivation, or elimination of pathogens in drinking water. This process is a necessary barrier to transmission of waterborne disease by microorganisms. Disinfection provides a large measure of public health protection at a relatively low cost. The accepted practice within the drinking water industry in the United States is chemical disinfection. The vast majority of U.S. water utilities use chemical disinfection with free chlorine, chloramines, ozone, chlorine dioxide, or combinations of these disinfectants. Of these, free chlorine is used most commonly. Until recently, disinfection by irradiation with ultraviolet (UV) light was practiced primarily within the wastewater industry; however, its use within the drinking water industry has lately been increasing. The USEPA has finalized a guidance manual for UV disinfection of drinking water.

It is common practice to regulate disinfection in water treatment plants by requiring that the drinking water meet CT value before it is delivered to the first customer. USEPA has adopted the use of CT values as a measure of the degree of disinfection achieved. CT values are the product of the disinfectant concentration, C (mg/L), and the contact time, T (minutes), at the point of disinfectant residual measurement. Disinfectant contact times used in calculating CT values are determined by field studies using tracer compounds. Contact time is defined as the T_{10} detention time, or the period during which 10 percent of the water entering a specific treatment unit (settling basin, clearwell, filter, etc.) has passed through the unit. This definition of T_{10} ensures that at least 90 percent of the water being treated remains in contact with the disinfectant for the reported length of time. The T_{10}/T_{theory} ratio is used as a measure of the short-circuiting which occurs within the basins. T_{theory} is the theoretical detention time which is calculated using the basin volume and a specific flow rate. These ratios range from 0 to 1, and as the value approaches 1, the basin experiences less short-circuiting and is simulating plug flow.

As directed by the Safe Drinking Water Act Amendments of 1986, USEPA has established criteria for regulation of five microbial contaminants in drinking water derived from surface supplies: *Giardia lamblia* cysts (*Giardia*), enteric (intestinal) viruses, *Legionella*, heterotrophic bacteria (HPC), and coliforms. USEPA has recognized that it is neither economically nor technically feasible to measure the concentrations of these contaminants in water on a regular basis. It has, therefore, promulgated treatment techniques that will result in their removal and/or inactivation, with primary focus on controlling *Giardia* cysts and enteric viruses (USEPA, 1991). When these two contaminants are effectively inactivated, the remaining three are also reduced to acceptable levels. The treatment techniques for the control of these microbial contaminants are specified in the Surface Water Treatment Rule (SWTR), so that a minimum of 99.9 percent (3 log) removal and/or inactivation is achieved for *Giardia* cysts and 99.99 percent (4 log) for enteric viruses (USEPA,1989). The proposed Long Term 2 Enhanced Surface Water Treatment Rule (LT2SWTR) does focus on *Cryptosporidium* inactivation. The USEPA is concerned with *Cryptosporidium* because it is resistant to chemical disinfection. *Cryptosporidium* has also been involved in several waterborne disease outbreaks.

Disinfection is usually discussed in terms of "log" inactivation; that is, reductions in the number of pathogens on a log scale. For instance, if a water sample initially contains 100 pathogens per liter, reducing this population to 10 pathogens per liter would be 1 log (90 percent) removal. Further reducing the pathogen population to 1 per liter would be 2 log (99 percent) removal, and 1 per 10 liters would be 3 log (99.9 percent) removal.

Credit is given for the physical removal of viruses and cysts through the filtration process. The number of log credits varies with the type of plant (conventional treatment, direct filtration, etc.). For example, the Surface Water Treatment Rule (SWTR) states that direct filtration systems typically receive a 2.0 log credit for *Giardia* removal or 1.0 log credit for virus removal. The remaining credit can be met with various disinfection technologies. Two such technologies that are discussed in length in this report are ozonation and UV irradiation.

VARIOUS CHEMICAL DISINFECTANTS

Free Chlorine

For years, water utilities have used chlorine for disinfection of drinking water. After the discovery of trihalomethanes (THMs) in drinking water, water utilities examined means of alternative treatment to reduce the use of chlorine, especially for pretreatment. Today, chlorine is still most commonly used for disinfection; however, many water treatment facilities convert free chlorine to chloramines before distribution, to reduce THM formation and to maintain a sustainable disinfectant residual in the distribution system.

Free chlorine is a relatively inexpensive disinfectant, and it has been very effective for the inactivation of many kinds of microorganisms, which contributed to its widespread use. However, the use of free chlorine does have several disadvantages. Its effectiveness is pH dependent. At pH above 7, hypochlorous acid (HOCl), the more powerful form of free chlorine, disassociates to form hypochlorite ion, OCl^-, a weaker disinfectant. Thus, as the pH increases above pH 7, disinfection with free chlorine becomes less effective. This has implications for distribution systems because of concerns regarding corrosion control and compliance with the Lead and Copper Rule. Free chlorine also has not been found to be as effective as chloramine for controlling biofilms in the distribution system. Another drawback to using free chlorine as a disinfectant is the formation of disinfection by-products (DBPs). At higher pH values, the reaction of chlorine with DBP precursors to form THMs is accelerated, and higher yields of THMs are realized. At lower pH values, THM formation is lessened, however the formation of some HAAs and other DBP is enhanced.

Free chlorine used alone has not proved to be an effective disinfectant for inactivating *Cryptosporidium*. Work by Finch et al. (1998) indicates that free chlorine followed by chloramination is more effective for inactivating *Cryptosporidium* than either free chlorine or chloramine used alone; however, data developed by Oppenheimer et al. (1998) seem to contradict these findings.

Finch and co-workers evaluated *Cryptosporidium* oocyst inactivation by free chlorine (Finch et al. 1994). Free chlorine alone was ineffective, attaining about 0.2 log inactivation after

240 minutes of contact at 15 mg/L (a CT of about 3600 mg * min/L) at pH 8.0 and 22° C. A free chlorine residual of 15 mg/L would cause formation of very high concentrations of DBPs in most waters.

Hirata (1999) evaluated the effects of free chlorine on the viability and animal infectivity of *Cryptosporidium parvum* oocysts. Chlorination experiments were carried out at pH 7 and 20°C. A 3-log reduction was observed at CT product of 2900 mg*min/L(48 hour contact with 1 mg/L chlorine) and estimated CT product for 1 log reduction in animal infectivity was 900 mg*min/L. Chlorine could be expected to reduce the infectivity of the *Cryptosporidium parvum* oocysts to some extent at practical chlorine contact times and concentrations, but only if the water being treated was low in DBP precursors. For waters having TOC concentrations above 3 to 4 mg/L, attaining a CT of 900 mg*min/L would be likely to cause excessively high concentrations of trihalomethanes or haloacetic acids. At least some of the difference between the results of Finch et al. (1994) and those by Hirata (1999) would be caused by the lower pH used by the latter research team.

Marinas et al. (1999a) found that the inactivation of *Cryptosporidium parvum* oocysts was strongly dependent on pH, using hypochlorous acid. In general, CT values for hypochlorous acid were approximately 1,000 times higher than those required to achieve the same inactivation efficiency with ozone. Temperature greatly affected the inactivation kinetics of free chlorine (Marinas et al 1999a).

Monochloramine

Chloramines come into wider use as a result of their ability to provide a degree of disinfection without substantial THM formation. Taste and odor control and maintaining a more stable residual in distribution systems are other benefits of chloramine use. If improperly managed, however, the use of chloramines can support bacterial growth in the distribution system.

Chloramination is accomplished by combining free chlorine with ammonia or !n ammonium salt, to form chloramine. Chloramine is not as strong as chlorine for disinfection, and it is not recommended by USEPA as a primary disinfectant. Chloramine does, however, form a persistent disinfectant residual, and is used by numerous water utilities for maintenance of a residual in the distribution system. Chloramine is slower to react with substances that adhere to the walls of water mains and thus has a better opportunity to penetrate tubercles and biofilms and kill resident bacteria.

Finch and co-workers evaluated the inactivation of *Cryptosporidium* oocyst by monochloramine (Finch et al. 1994). Monochloramine alone was ineffective, attaining about 0.2 log-units of inactivation after 8 hours of contact at 5 mg/L (a CT of 2400 mg*min/L) at pH 8.0 and a temperature of 22°C. Monochloramine alone is not a very effective disinfectant against *Cryptosporidium*. Even though chloramines are not considered strong, primary disinfectants, their use in sequential disinfection for inactivation of *Cryptosporidium*, where a strong disinfectant such as free chlorine, chlorine dioxide, or ozone is followed by prolonged contact with chloramine, has shown promise.

Efficacy of chloramine alone was evaluated by Driedger, Rennecker, and Marinas (1999) as part of an investigation of the synergistic effects of sequential ozone and chloramine disinfection. Assay by a modified in vitro excystation method was reported by the authors to give results consistent with an animal infectivity assay. The results of Driedger et al. (1999) for CT in the range of 2000 to 3000 mg * min/L are similar to those of Finch et al. (1994) for chloramine alone at a CT of about 2400 mg*min/l, i.e. less than 0.5-$\log_{10}$ inactivation. The 2.3 $\log_{10}$ kill reported by Finch et al (1994) at a CT of 7200 mg*min/L, pH 8.0, and 22°C compares to the results seen by Driedger, Rennecker, and Marinas at a CT of 12,000 mg*min/L. Qualitatively, both sets of data indicate that chloramine alone would not be very effective against *Cryptosporidium*.

Chlorine Dioxide

Chlorine dioxide (ClO_2) is a very effective disinfectant. Most often, it is used as a substitute for chlorine as a preoxidant and is applied to the raw water. The practical ClO_2 dosage applied during water treatment is restricted considerably by Maximum Contaminant Levels (MCLs) imposed in the D/DBP Rule on ClO_2 (0.8 mg/L) and its major by-product, chlorite ion ClO_2^-, (1.0 mg/L). One major change in the Stage I D/DBP Rule is the increase in the chlorite MCL (from 0.08 mg/L as originally proposed by the USEPA). One advantage to using chlorine dioxide is the reduced DBP formation. As a result, utilities that use or have initiated the use of ClO_2 meet at least part of the overall oxidant demand of the water supply without forming DBPs. Nevertheless, utilities seldom use ClO_2 primarily for its disinfecting properties.

Finch et al. (1998) reported a 1.5 log inactivation for *Cryptosporidium* when chlorine dioxide at a concentration of 0.9 mg/L was applied for 45 minutes, followed by monochloramine at 0.6 mg/L for 30 minutes. Ruffell, Rennecker, and Marinas (1998) reported their research on *Cryptosporidium* inactivation. In contrast to Finch's work, this researcher used excystation rather than animal infectivity to determine inactivation. They reported that a CT of 60 mg*min/L could attain in excess of 1 log inactivation at pH 8 and 20°C. At these conditions, a CT of 75 mg*min/L yielded 2 log inactivation.

There are some disadvantages associated with ClO_2. The oxidant is a solution at temperatures greater than 11°C and is easily volatilized if the generator effluent is added to the rapid mixer and is exposed to the atmosphere during the contact period. Because solubility of ClO_2 decreases at higher temperatures, a utility can expect greater losses of the gas from uncovered treatment basins during warmer weather. Odor complaints from plant personnel and neighboring residents can increase when the water temperature is higher. Inactivation kinetics of *Cryptosporidium* oocysts by chlorine dioxide have been investigated (Ruffell, Rennecker and Marinas, 1998) using a modified in vitro excystation technique reported by the authors to give results consistent with animal infectivity. Testing was done at 20°C, and at pH 6, 8, and 10. The CT values that gave 2-log inactivation were 120, 75, and 60 mg*min/L. respectively. The authors indicated that the apparently greater efficacy of chlorine dioxide at pH 10 was not expected based on chemistry of chlorine dioxide.

A study by Marinas and coworkers (1999a) found that the inactivation of *Cryptosporidium parvum* oocysts with chlorine dioxide was independent of pH in the range of 6 to 10. The same study also reported the CT requirements were approximately 50 times greater than those needed to achieve the same inactivation with ozone. Temperature greatly affected the inactivation kinetics of chlorine dioxide, with cold water requiring greater CT values in this study.

A comparison of assay methods was undertaken during research into inactivation of *Cryptosporidium* oocysts by chlorine dioxide (Owens et al. 1999). Inactivation was measured by neonatal mouse infectivity, by a modified *in vitro* excystation method described by Rennecker et al. (1999), and by a human cell culture assay used in another laboratory (Slifko et al. 1996). The authors reported that the cell culture infectivity and the modified excystation procedure understated the extent of inactivation as measured by animal infectivity. On the basis of their data, the difference was about 1/2 $\log_{10}$.

OZONE DISINFECTION

Ozone is a powerful disinfectant, which can be used to inactivated *Giardia lamblia* and *Cryptosporidium*. Based on the disinfection criteria used by the USEPA, ozone is 100 to 300 times more effective than chlorine in inactivating *Giardia* cysts. The use of ozone reduces the formation of chlorinated by-products such as trihalomethanes (THMs) and other disinfection by-products (DBPs) produced when chlorine reacts with the naturally occurring organic matter (NOM) contained in many water sources.

CT Concept for *Cryptosporidium, Giardia* and Virus Inactivation

The USEPA has published CT values for disinfection with ozone in the Surface Water Treatment Manual (SWTR) manual as shown in Tables 2.1 and 2.2. CT values are dependent on water temperature when ozone is the primary disinfectant and increase as water temperature drops.

The disinfection CT requirements for the ozone contact basins are different from those for chlorine detention basins. Since ozone can be added to the contactor in multiple locations, the CT values are calculated for each cell (instead of only at the end of the basin, as with chlorine contact basins). Typical T_{10}/T_{theory} ratios for ozone contact basins range from 0.5 to 0.7, while T_{10}/T_{theory} ratios for detention basins vary with different basin designs and flow characteristics.

Special consideration is given to the first cell of the ozone contactor. For the first cell of the contactor, CT values are not calculated using USEPA's conservative method. Instead, flat log credits are granted under certain conditions. For example, if a 0.1 mg/L ozone residual is maintained in the outlet of the first cell, a 1 log virus inactivation credit is granted. If a 0.3 mg/L ozone residual is maintained in the outlet of the first cell, a 1 log virus and a 0.5 log *Giardia* inactivation credit is granted. A summary of the log removal credits for the first cell of the ozone contactor is outlined in Table 2.3. Based on the CT values listed in Tables 2.1 and 2.2, the maintenance of a 0.1 mg/L dissolved ozone residual in the effluent of cell 1 substantially reduces the CT required from the remaining cells. Increasing the residual to a concentration greater than 0.3 mg/L offers only minimal benefits in reducing the CT required from the remaining cells.

Table 2.1
CT values for *Giardia* cyst inactivation by ozone

Log inactivation	Temperature (°C)					
	1	5	10	15	20	25
0.5	0.48	0.32	0.23	0.16	0.12	0.08
1.0	0.97	0.63	0.48	0.32	0.24	0.16
1.5	1.5	0.95	0.72	0.48	0.36	0.24
2.0	1.9	1.3	0.95	0.63	0.48	0.32
2.5	2.4	1.6	1.2	0.79	0.60	0.40
3.0	2.9	1.9	1.43	0.95	0.72	0.48

Source: USEPA Surface Water Treatment Rule 1990.

Table 2.2
CT values for virus inactivation by ozone*

Log inactivation	Temperature (°C)					
	1	5	10	15	20	25
1.0	0.45	0.30	0.25	0.15	0.125	0.075
1.5	0.59	0.39	0.33	0.2	0.16	0.10
2.0	0.9	0.6	0.5	0.3	0.25	0.15
2.5	1.04	0.69	0.58	0.34	0.29	0.17
3.0	1.4	0.9	0.8	0.5	0.4	0.25
4.0	1.8	1.2	1.0	0.6	0.5	0.3

Source: USEPA Surface Water Treatment Rule, 1990.

* Values at 1.0, 1.5, and 2.5 log inactivation are based on using the kinetic "k" values for virus inactivation listed in the appendix O of the SWTR

Table 2.3
Inactivation credits for ozone contact basins

Ozone residual at outlet of cell 1 (mg/L)	Inactivation/removal credit
< 0.1	**None**
$\geq$ 0.1 but < 0.3	1.0 log virus
$\geq$ 0.3	1.0 log virus and 0.5 log *Giardia*

Source: USEPA Surface Water Treatment Rule, 1990.

The design of the ozone contactor dictates the method for calculating CT in any subsequent chamber of the ozone contactor. USEPA recommends that the characteristic dissolved ozone concentration for each cell be determined by directly measuring dissolved ozone at several points in the cell and computing the average concentration from these measurements. However, USEPA also provides a simplified, albeit conservative, method of calculating the characteristic concentration, C, for use in calculating CT values that uses the effluent ozone concentration from each cell. For a contactor consisting of 8 cells, with ozone being applied countercurrent to the liquid flow in cell 1, flat credits would be granted for cell 1. For the remaining cells (2 through 8), the dissolved ozone residual measured at the cell outlet is used directly in calculating CT values. While the T_{10} time is usually determined for the entire ozone contactor, the T_{10} of each cell is used to calculate CT values. Using the conservative method, the CT value would be equal to the following:

$$CT = SCF*[(C_{cell\ 2}*T_{cell\ 2}) + (C_{cell\ 3}*T_{cell\ 3}) + \ldots + (C_{cell\ 7}*T_{cell\ 7}) + (C_{cell\ 8}*T_{cell\ 8})] \qquad (2.1)$$

where: SCF = T_{10}/T_{theory} (short-circuiting factor - typically 0.50 - 0.65)
$T_{cell\ x}$ = theoretical detention time for Cell X (minutes)
$C_{cell\ x}$ = effluent dissolved ozone concentration from Cell X (mg/l)

Using this equation, it can be concluded that the addition of more sampling points results in a more accurate determination of the actual disinfection environment in the ozone contactor. As an extreme condition, sampling points could be installed one after the other along the entire length of the contactor to arrive at the most accurate calculation of CT. However, if the rate of ozone decay follows first-order reaction kinetics, the ozone residual at any point in the contactor can be calculated (Coffey. 1994). By measuring the ozone residual at two points in the contactor where the detention time between those points is known, the ozone decay rate, k, can be calculated. With a constant decay rate and a known initial ozone residual, the log-integration approach can be used to calculate the CT value. The equation used to calculate CT based on the "log-integration" method would be as follows:

$$CT = (T_{10}/T_{theory}) \times (C_0) \times (e^{(kt)} - 1) / k \qquad (2.2)$$

This technique can result in a reduced number of sampling locations while providing a higher, more accurate CT value than the "conservative correlation" method, because the log-integration method provides a more accurate representation of the area under the curve of dissolved ozone residual versus detention time. This is illustrated on Figure 2.1. Both methods were used to calculate CT values using ozone demand and decay data from previous testing at another facility.

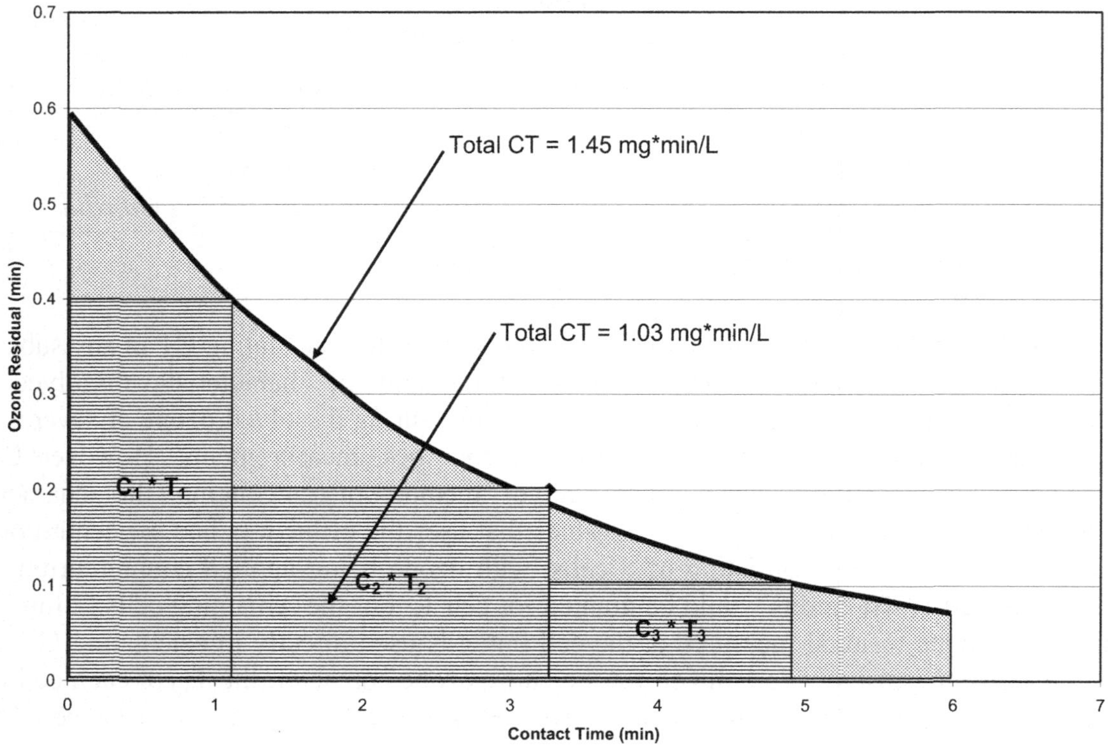

Figure 2.1 Comparison of conservative correlation and log-integration

The log-integration method was found to result in a CT value between 10 and 25 percent higher than the conservative method.

In a conventional ozone contactor, the ozone residual measured at the end of the first countercurrent cell has the highest concentration in the cell. USEPA has recommended that flat credits of 1-log inactivation of viruses and 0.5-log inactivation of *Giardia* cysts be granted for achieving ozone residuals of 0.1 and 0.3 mg/L at the end of the first cell, respectively. No information is available on a flat credit for *Cryptosporidium*. It may be necessary to profile the dissolved ozone residual in the first cell to obtain disinfection credit for *Cryptosporidium* in the future. If so, the log-integration equation would be altered to take into account the characteristic ozone residual in the first cell, as shown in the following equation.

$$CT = (SCF)*[(C_{characteristic\ 1} \times T_{cell1}) + (C_0*(e^{(kt)} - 1)\ )/\ k] \quad (2.3)$$

The USEPA recently concluded regulatory negotiations, which may require some form of *Cryptosporidium* inactivation, dependent on the raw water quality and the type and efficacy of treatment utilized. The Long Term 1 Enhanced Surface Water Treatment Rule (LT1ESWTR) established a 2-log removal criterion for *Cryptosporidium*, and the Long Term 2 Enhanced Surface Water Treatment Rule (LT2ESWTR) will set further removal/inactivation requirements. Some systems will need to document additional removal efficiencies and others will need to provide inactivation. Systems with inactivation requirements will normally need to use alternative disinfection techniques, as free chlorine and chloramines are largely ineffective against this organism. EPA considers ozone and UV to be effective for disinfection.

The LT2SWTR proposes two methods for determining CT values. One alternative is to measure the initial ozone residual and final ozone residual and calculate a geometric mean CT. The second alternative is measuring the final ozone residual and using only that residual for calculation of CT. This alternative provides an option of a site specific investigation of the plant to determine log inactivation. This would require approval by the state. If USEPA adopts a CT approach for *Cryptosporidium*, then the approach to calculating CT described in the preceding section would be appropriate. However, other approaches have been other approaches proposed to calculate log-inactivation of *Cryptosporidium* that would require alternate calculations of ozone efficacy. A table consisting of mean CT values for Cryptosporidium inactivation by ozone as a function of water temperature was proposed in the LT2SWTR. It is based on a predictive equation created to calculate a mean CT value for a composite set of data (USEPA 2001).

OZONE DISINFECTION EFFECTIVENESS

Extensive efforts have been made to determine the effectiveness of ozone against pathogens. The pathogens of concern in water treatment are bacteria, viruses, and protozoans. A study by McNair and Lesher (1963) postulated that the introduction of ozone into the cell wall of the bacteria or the membrane of the bacteria will produce lysis of the cells, thereby inactivating the bacteria. This hypothesis can most likely be applied to the inactivation of other microorganisms in water treatment. The literature reviewed in the following sections is the result of various ozone applications, including bench-scale studies, pilot-scale studies and full-scale studies. The data of the studies suggest the variability in ozone efficiency as a result of due to varying water quality characteristics, (temperature and pH) and ozone design parameters (ozone dose and CT values).

Bacteria and Viruses

Ozone is very effective against bacteria and viruses. A study performed by Herbold, Flehmig, and Botzenhart (1989) showed that *E. coli* was more sensitive to ozonation at low doses of ozone (0.10 mg/L); however, it is also temperature dependent because as the temperature increased the effectiveness decreased (Herbold et al., 1989). Other bacteria such as *Staphylococcus* and *Streptococcus*, *Bacillus* and *Mycobacteria* are more resistant (Langlaid et al. Eds., 1991). Coffey et al. (1999) studied ozone disinfection using *Bacillus* spores. During these tests, the pH and temperature of the system were 6.5°C and 10°C. A CT of 7.7 mg*min/L was required to achieve a 1-log inactivation of the *Bacillus* spores.

In the study by Herbold, Flehmig, and Botzenhart (1989)., the authors compared ozone inactivation of Hepatitis A virus and Poliovirus 1. The results of this study indicated that the Poliovirus 1 was more sensitive to ozone than the Hepatitis A virus at 20°C at an ozone concentration of 0.10 mg/L.

Giardia

Ozone is a much more effective disinfectant than chlorine or chlorine dioxide for the inactivation of *Giardia*. Finch and Black (1993) stated that the ozone dose required to achieve 2-log inactivation of *Giardia lamblia* at 22°C was greater than what was previously reported. However, the indicator for inactivation in previous studies was *in vitro* excystation (Sauch et al., 1991) and the Finch and Black study used animal infectivity. Their study indicated that a 2-log inactivation required a CT of 1.2 mg*min/L. It was also shown that to achieve 3-log inactivation of *Giardia* at 22°C, a CT of 1.6 mg*min/L was required, based on an integrated CT calculation. The USEPA CT value was estimated from extrapolation of the simple Chick-Watson kinetic model.

Cryptosporidium

Kanjo et al. (2000) evaluated inactivation of *Cryptosporidium muris* oocysts treated with 0.5 mg/L of ozone for 0, 10, 30, and 50 minutes at room temperature. Inactivation ratios derived from animal infectivity tests for each treated group were 0 percent, 99 percent, 99.9 percent, and over 99.9 percent, (0, 2 log, 3 log, and > 3 log, respectively). The required CT values were about 3 and 5 to 7 mg*min/L for 99 percent and 99.9 percent inactivation, respectively. Inactivation ratios for *Cryptosporidium* at 5°C decreased, and the required CT values were around 9 and 20 mg*min/L for 99 percent and 99.9 percent, respectively.

Gigliotti, States, and Casson (1999) used bench-scale and pilot-scale testing to evaluate *Cryptosporidium* inactivation with ozone. Using *in vitro* excystation, bench scale testing produced a 1.2-log reduction with a CT of 30.0 mg*min/L at 3.5°C and 11.0 mg*min/L at 21°C. Pilot-scale testing was conducted using two 10.75 feet long, 4 inch diameter glass contact chambers in series and a Mazzei injector for ozone addition. Fluoride tracer studies were conducted to verify the detention time of the system. Pilot-scale testing produced an inactivation range from 0.1 log with a CT of 6.4 mg*min/L at 11°C to 0.5 log with a CT of 18.2 mg*min/L at 3°C, as measured by *in vitro* excystation.

A literature study by Finch et al. (2001) suggests that any comparison involving *Cryptosporidium* inactivation is a subjective comparison. The authors reasons were as follows: "different strains of parasites have been used, parasite are propagated using different methods, parasites are stored under different conditions, and there is a potential disparity in viability measurements".

Owens et al. (1994) conducted pilot plant studies of ozone inactivation using an ozone contactor with a water depth of 2.65 m and a diameter of 0.15 m. Test conditions were pH of 7.5 to 8.5, temperature 22 to 25°C, and TOC concentration ranging from 1.6 to 2.1 mg/L. Inactivation was evaluated by animal infectivity. Under the above-stated conditions, they attained a 2.0 log inactivation of *Cryptosporidium* with a CT value of 5.5 mg*min/L.

Hirata (1999) evaluated the effects of ozone on the viability and animal infectivity of *Cryptosporidium* oocyst. Ozonation experiments were carried out at pH 7 and 20°C. The CT product of ozone required for 2-log reduction in animal infectivity of *Cryptosporidium* oocysts was calculated at approximately 6 mg*min/L.

Due to the difficulty of animal infectivity assays, Motoyama et al. (1999) examined the CT values required for 1 log and 2 log inactivation of *Cryptosporidium* with ozone using the excystation and fluorescent vital dyes 4′,6-Diamino-2-phenylindole and propidium iodide (DAPI /PI) staining tests. The CT values for 1 log and 2 log inactivation using the excystation method were 6.3 to 7.6 mg*min/l and 10 to 12 mg*min/L, and for the DAPI/PI method were 7.5 to 12 mg*min/L and 16 to 24 mg*min/L, respectively.

IMPACTS OF WATER QUALITY ON OZONE EFFECTIVENESS

Temperature and pH

Application of ozone for effective disinfection requires knowledge of the characteristics of the water to be treated. Procedures for evaluation of ozone demand and decay have been described by Hulsey et al. (1997). Eliminating the ozone residual after disinfection is necessary when filters are enclosed in a building and ozone is applied prior to filtration. Hulsey et al. (1999) described ozone process design and ways of dealing with excess residual ozone when disinfection has been accomplished.

Oppenheimer et al. (1999) concluded that ozone is an effective disinfectant for inactivation of *Cryptosporidium*, but the CT product to achieve a target level of inactivation may be significantly impacted by the temperature, the pH, and the turbidity of the water. Inactivation by ozone is significantly enhanced if the pH of the water is 6.5 or lower and if the turbidity of the water is low. The influence of water temperature on ozone inactivation kinetics for *Cryptosporidium* appears to be greater than the effect previously demonstrated for *Giardia lamblia* and utilized in establishing the CT tables for *Giardia* inactivation in the Surface Water Treatment Rule Guidance Manual.

The effect of cold water on inactivation of oocysts by ozone at pH 6 was very significant (Li, Finch, and Belosevic, 1999). At 1°C, an ozone dosage and contact time CT value ranging from 45 to 54 mg*min/L gave about 1.4 log inactivation, but at 22°C the CT values in the range from 3.6 to 4.5 mg*min/L resulted in 1.5 log units of inactivation. In general, when water temperatures are very low, disinfection is more difficult. These data give an indication of the degree of difficulty that would be encountered in managing disinfection when the target organism is *Cryptosporidium*.

A study by Marinas et al. (1999a) found that inactivation of *Cryptosporidium parvum* oocysts with ozone is independent from pH in the range of 6 to 10. Temperature greatly affected the inactivation kinetics of all the disinfectants. For example, the CT required for 3 log (99.9 percent) inactivation with ozone increased from 2 mg*min/L at 30°C to nearly 50 mg*min/L at 5°C.

ULTRAVIOLET RADIATION

UV Disinfection Overview

Background

It was Wheatstone who, in 1835, first documented the intense light emitted when vaporizing mercury with an electric arc. Later, in 1901, the mercury vapor lamp was commercialized as a method of UV generation. Built in Marseilles, France in 1910, the first UV disinfection system was unreliable and complicated to operate due to the state of UV technology at that time. Because of the competition from the use of lower-costing chlorine for disinfection purposes, UV development lagged behind for several decades. In the 1950's, there began a renewed interest in UV disinfection technology due to advances in UV lamps and electronics as well as growing taste and odor concerns with using chlorine (DeMers and Renner 1992).

At present, virtually all of the full-scale drinking water applications of UV disinfection for public water supplies are located in Europe. Specific European applications for UV include the following (Kruithof, van der Leer, and Hijnen 1992):

- In groundwater treatment scenarios for destruction of *E. coli* and *Aeromonas* bacteria
- In bank filtration applications to reduce colony counts after carbon filters
- As a replacement for post-chlorination

It has been the Dutch experience that the effluent colony counts and the level of assimilable organic carbon (AOC) will determine the post-disinfection requirements. In waters with high levels of AOC, a secondary disinfectant that maintains a residual is used. Because of its effectiveness at destroying colony forming bacteria, UV disinfection is employed in situations where the effluent colony counts are high and the AOC is low (Kruithof, van der Leer, and Hijnen 1992). This reduces the number of viable bacteria which can proliferate in the distribution system. Table 2.4 details some of the European experience with UV disinfection. By the early 1990's, there were over 2000 drinking water installations in Europe using UV irradiation (DeMers and Renner 1992). In addition to municipal drinking water disinfection, UV is also approved for water disinfection on ships (Witherell, Solomon, and Stone 1979).

Application of UV disinfection in the United States has followed a different path. UV disinfection has gained widespread use for wastewater throughout the U.S. since it does not involve the transportation, storage and handling of regulated chemicals like chlorine, nor does it leave a residual in the treated effluent that could contribute to whole effluent toxicity. More recently, interest in using UV for water reuse applications has increased due to results of promising pilot and full-scale tests (Montgomery Watson, 1994). However, use of UV in U.S. drinking water applications has been limited primarily to small point of entry or point of use systems treating individual ground water supplies.

Table 2.4
European UV disinfection facilities

Country	First application	Installations in 1985	Capacity (MGD)
Switzerland	1955	500	2.5
Austria	1955	600	3.2
Norway	1975	400	6.3

Source: Kruithof, van der Leer, and Hijnen 1992.

The three primary reasons that use of UV for drinking water treatment in the U.S. has been limited are: 1) the fact that UV does not leave a residual which could serve as protection for the distribution system; 2) the fact that UV was not considered effective for inactivation of *Giardia lamblia* cysts and *Cryptosporidium parvum* oocysts; and 3) the fact that most ground water systems in the U.S. have not been required to provide primary disinfection for bacteria and viruses.

UV Disinfection Fundamentals

UV Light

UV light is electromagnetic energy that is located in the electromagnetic spectrum at wavelengths between those of X-rays and visible light. Figure 2.2 shows that UV is further broken down into UV-A, UV-B, UV-C, and Vacuum UV. These types of light reside from about 40 to 400 nanometers (nm) in wavelength. The UV light effective in inactivating bacteria and virus resides in the UV-B and UV-C ranges of the spectrum (200 to 310 nm). Maximum effectiveness for inactivation of most bacteria and virus species occurs around 265 nm (Black & Veatch 1995).

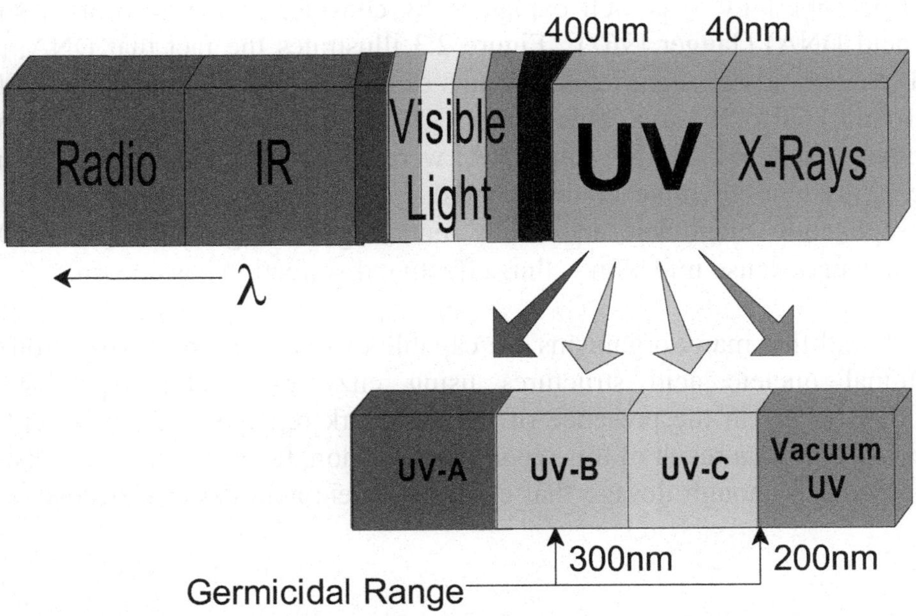

Figure 2.2 Electromagnetic Spectrum

In order to develop an understanding of UV light energy, it is useful to review several key principles. The relationship between frequency and wavelength of electromagnetic energy is defined by Equation 2.4:

$$\nu = \frac{c}{\lambda} \qquad (2.4)$$

Where c = speed of light ($3x10^8$ m/sec)
ν= frequency (Hz or cycles/sec))
λ = wavelength (m)

The energy contained in the fundamental unit of electromagnetic radiation (photon) at a given frequency is defined by Equation 2.5 (Halliday and Resnick 1988):

$$E = h\nu = \frac{hc}{\lambda} \qquad (2.5)$$

Where E = Energy (Joules)
h = Planck's Constant (6.63×10^{-34} J-sec)

The quantity of light can also be referenced in terms of an Einstein. One Einstein equals one mole (6.022×10^{23}) of photons at the wavelength considered. Equations 2.4 and 2.5 allow the relationship between wavelength, Einstein's of photons and energy to be determined. For example, at a wavelength of 253.7 nm, 1 Joule (or 9 W-sec) of energy is equivalent to 2.12 Einstein.

Germicidal Action of UV Irradiation

It is a fundamental premise of photochemistry that reactions cannot occur if the reactant species does not absorb energy in the spectral range of the incident light. Therefore, for UV to be an effective disinfectant some critical biochemical moiety of the given organism must absorb energy in the appropriate range. The photochemical interactions between UV light and living organisms, termed photobiology, was well explained by considering the adsorption spectra of deoxyribonucleic acid DNA) (Jagger,1967). Figure 2.3 illustrates the fact that DNA, and thus living cells such as *E. coli*, adsorb significant amounts of light energy in the range of 200 to 300 nm with a peak around 260 to 265 nm (Muelemans,1987). Further investigation indicated that thymine bases on the nucleic acids (DNA and RNA) were particularly reactive to UV light and would form dimers (thymine=thymine double bonds). These dimers then chemically inhibit transcription and replication of nucleic acids thus rendering the organism sterile. Once the organism is sterile it cannot cause infection -- thus effective disinfection has occurred.

It has been found that many organisms are capable of repairing the thymine dimers and restoring the original nucleic acid structures using enzymes. This repair is termed Photoreactivation if it occurs in the presence of light or Dark repair if light is absent (Jagger 1967; Stewart et al. 1993). As a result of this repair phenomenon, the strategy in UV disinfection has been to provide a high enough dosage that enough nucleic acid damage occurs to prevent effective repair.

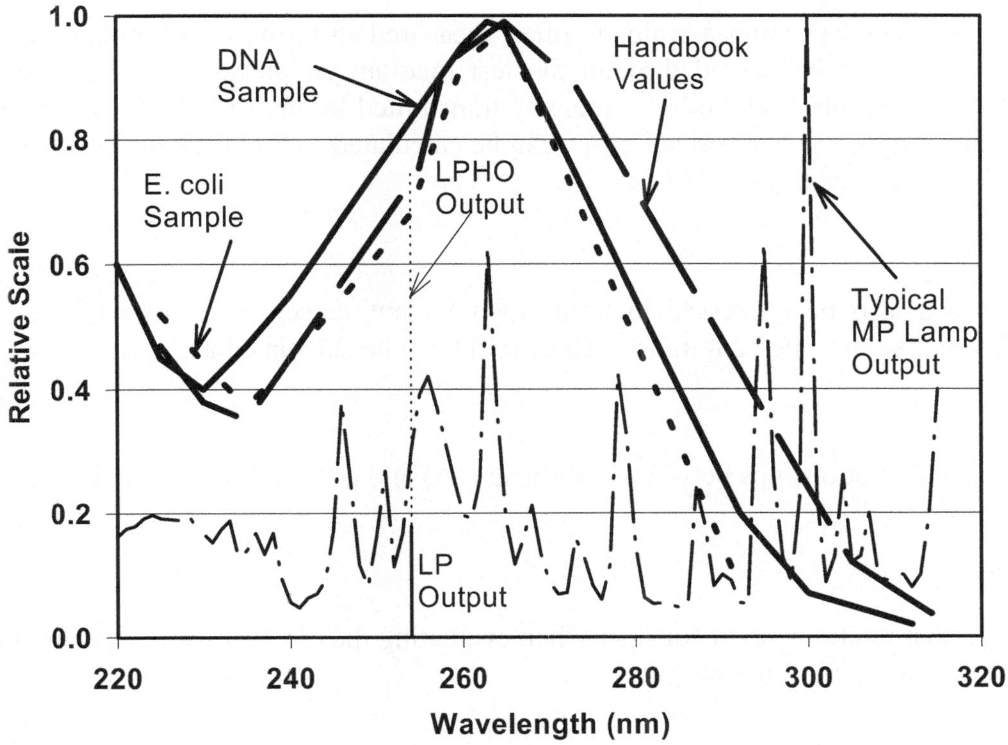

Figure 2.3 Absorbance spectra for DNA and E. coli with corresponding UV lamp outputs

This classical explanation of the germicidal mechanisms of UV irradiation holds extremely well in cases where continuous wave, monochromatic, moderate to low energy UV light is used – which has been the standard case for the past thirty years. More recently however, continuous or pulsed wave, polychromatic, high intensity UV lamp technologies have been developed and applied to water and wastewater treatment. Continuous wave, medium pressure mercury vapor lamps are becoming increasingly popular in higher flow water and wastewater applications due to overall cost savings. Pulsed UV technology which relies upon a high energy xenon arc (or flash) lamp have shown promising results in recent drinking water treatment research at the bench, pilot and full-scale (LaFrenz 1997; Mofidi et al. 1998) Medium pressure lamps emit significant amounts of energy at 222.5, 253.7, 265 and 296.5 nm (with lower emissions at several other wavelengths in the germicidal range). Pulsed UV lamps emit high energy bursts (pulses) of polychromatic light in the range of 185 to 800 nm.

It is theorized that these higher intensity, polychromatic emissions may do significantly more biochemical damage (similar to that of ionizing radiation) to not only nucleic acids but to cellular proteins and enzymes (Rubin, 1981; Budowsky et al., 1992). Related work by other researchers (Craggs et al., 1994; von Sonntag, 1986) suggests that these high intensity sources may enhance the formation of secondary radical reactions giving rise to germicidal chemical species such as hydroxyl radicals, ozone or hydrogen peroxide.

Determination of UV Attenuation and Absorbance Coefficient α

UV attenuation through a water is often measured in terms of absorbance for a fixed pathlength. Light can be passed through the test medium at intensity I_o at the desired UV wavelength (e.g. 254 nm) and the light intensity transmitted by the sample, I_t, can be measured for a fixed path length. The absorbance (A), can be calculated as the difference of these two log transformed intensity readings as follows:

$$A = \log_{10} \frac{I_0}{I_t} \tag{2.6}$$

UV demand can also be expressed in terms of a transmittance, or fraction of light passing through a given medium. Percent transmittance (%T) can be calculated as:

$$\%T = \frac{I_t}{I_0} \times 100 \tag{2.7}$$

The mathematical relationships between absorbance (A) and percent transmittance (%T) are:

$$\%T = 100 \times 10^{-A} \tag{2.8}$$

$$A = \log_{10} \frac{1}{T} \tag{2.9}$$

Another important relationship to consider when evaluating the effects of water quality matrix on UV irradiation is the Beer-Lambert Law:

$$A = \varepsilon\, L\, C \tag{2.10}$$

Where:

ε = the molar extinction coefficient, or molar absorptivity (volume/mole/length)
L = distance light passes through the homogeneous absorbing chemical – pathlength (length)
C = concentration of homogeneous absorbing chemical (mole/volume)

The water and wastewater industry often uses a modified form of the Beer-Lambert Law because it is impossible to specify a single homogeneous chemical concentration in water and wastewater matrices. Therefore, an absorbance coefficient α is used to represent the $\Sigma\ (\varepsilon_i C_i)$. The resulting expression is:

$$A = \alpha\, L \tag{2.11}$$

Combining equations 2.6 and 2.11 we find that the overall attenuation of the UV light (U), which is an indicator of interference with UV disinfection, can be expressed as:

$$U = (I_t / I_o) = \exp\,[\,-\,\alpha\, L] \tag{2.12}$$

Examination of equations 2.6 and 2.11 indicates that:

$$\alpha = (A/L)\ \ln(10) = 2.303\ (A/L) \tag{2.13}$$

In water and wastewater, A/L (also called UV_{254}) is measured using Standard Methods 5910 (APHA 1995) or equivalent and thus α can be easily calculated.

UV Dosage Determination

Because UV is a physical rather than chemical disinfectant, measurement of UV system performance variables (dose, irradiance, or relative lamp output) must occur within the reactor on a routine or continuous basis. This is different from the measurement of chemical disinfectants where a sample of the flow can be taken and processed. UV dose is expressed in a similar manner to that of conventional chemical disinfectants. It is a product of the UV irradiance (often called intensity and typically expressed in mW/cm^2) and the contact time with the UV source (sec). The result is commonly known as the Irradiance *X* Time product or IT with units of $mW\text{-}sec/cm^2$ (or J/m^2). This is analogous to the Concentration *X* Time, or CT product used with conventional chemical disinfectants.

Values for contact time in flowing systems are obtained through tracer studies in which the resulting hydraulic detention time (Θ_H) is determined. For a conservative estimate of Θ_H, the time at which 90 percent of the water is still detained (T_{10}) is often used. In drinking water applications it is often difficult to obtain accurate tracer data because the contact time is often 5 to 10 seconds. In these cases, Θ_H is either calculated from the theoretical reactor volume divided by the flowrate (V/Q) or from a mathematical model.

Although the concept of IT is simple, its practical determination in a full-scale UV reactor on a routine or continuous basis is difficult. In all full-scale UV reactors there is a distribution of irradiance and a distribution of fluid velocities (residence times). Therefore, to accurately assess IT or dosage it is necessary to adequately quantify both of these distributions. The task becomes even more difficult for newer UV light sources that emit polychromatic light thus requiring an integrated irradiance determination (Linden and Darby 1997). Presently, there are four techniques for determining UV dosage: biological, electronic, chemical and mathematical. However, one must use caution because depending upon technique chosen one can get very different answers for the applied UV dosage.

Biological UV Dosage Determination

First introduced by Qualls and Johnson (1983), these biological techniques intrinsically allow accounting for more factors that influence UV dose (Harris et al. 1987a). These tests are usually performed in the same water that is treated by an on-line UV system or proposed system so that the water matrix effects are inherent in the test results. However, laboratory irradiation of cultures is independent of hydrodynamic effects such as short circuiting or dispersion that can be present in flowing systems. Bioassay procedures examine a microorganism's response when exposed to a specific UV dose, known as the organism's UV susceptibility. For most microorganisms, there exists a first order relationship between UV dose and log inactivation.

One particular bioassay method used in the wastewater industry, National Sanitation Foundation Standard 55 (NSF/ANSI,1991), uses *Bacillus subtilis* spores. Under controlled laboratory conditions, early experiments showed that a 4 log inactivation of spores was achieved at a UV dose of about 31 $mW\text{-}sec/cm^2$ (Qualls and Johnson 1983). Because drinking water applications will likely require a slightly higher UV dose, obtaining a sufficiently large initial population of spores to measure substantial inactivation is a problem. For this reason, other organisms that are more resistant to UV may be required to measure these higher UV doses. One such organism is MS-2 Coliphage, which is an F-specific single-stranded RNA virus. MS-2 is about 20 nm in diameter and can be used as a viral surrogate (Braunstein et al. 1996).

MS-2 has been shown to exhibit a first order UV inactivation and can be prepared in sufficiently high titers such that only a small test volume is required (Havelaar 1991). The method for detection is simple, rapid, and reproducible and results can be obtained in as little as 6 hours (Wilson et al. 1992). These relatively quick results can give utilities the opportunity to perform system optimization and monitor system performance by having response data that is closer to real-time than many other biological tests.

Before the dose of a UV system can be determined by the bioassay method, it is first necessary to generate a calibration curve relating the UV dose to log inactivation of the test organism. This can be done in the laboratory by irradiating a pure culture of organisms with a constant UV intensity in the test water matrix. Laboratory UV irradiation typically requires a beam collimator tube to absorb and UV energy that is not perpendicular to the sample so that a uniform intensity distribution is created around the sample. Multiple irradiation times should be performed with organisms enumerated both before and after irradiation. Laboratory irradiation of cultures allows strict control of UV intensity and irradiation time without the interference of hydraulic effects encountered with flowing systems.

Once a linear relationship between UV dose and the log inactivation of an organism is established, UV doses for full-scale operating systems can be tested by spiking the test organisms into a system and measuring log inactivation. Then by taking this log inactivation and referring back to the dose-response calibration curve generated in the laboratory, an average UV dose can be determined from the observed organism inactivation. The calibration curve should be prepared for the full range of test doses and not be extrapolated due to the existence of tailing caused by particulate matter, clumping, and highly resistant organisms (Braunstein et al. 1996). If desired, one can back calculate the average UV reactor irradiance by taking the bioassay determined UV dose and dividing by the theoretical or measured hydraulic detention time (Θ_H or T_{10}) of the UV reactor.

The bioassay methods are extremely useful in selecting UV dosages during pilot testing and design and may be used periodically to verify full-scale UV reactor performance, but they are time consuming and costly so they do not provide an adequate method for routine (real-time) UV dosage determination.

Electronic UV Dosage Determination

The irradiance of a UV light source can be determined electronically using a radiometer which generally consists of a wavelength specific light sensor and the necessary electronics to convert the signal to a readout in units of mW/cm^2 (or J/m^2). The corresponding UV dosage can then be determined by taking the radiometer or sensor reading and multiplying by the hydraulic detention time. This method has become the standard practice in most full-scale UV reactor applications since it can provide simple, cost-effective routine monitoring. However, several problems exist.

The first difficulty is determining the required number and placement of sensors in a full-scale reactor to adequately quantify the UV irradiance in the reactor. Actual irradiance is strongly influenced by lamp and reactor geometry, reflection and water quality matrix effects. The second difficulty is the long term reliability of UV sensors. UV sensors are commonly found to foul, fail electronically or be degraded (blinded or fused) by continuous exposure to UV light. Six commercially available UV sensors were used during ongoing research by projects at UNH-ERG (Malley et al. 1996; Snicer et al. 1998; Malley et al. 1998). Sensors #1 and #2 failed (produced erratic and meaningless results) almost immediately after installation due to electronics problems. Sensors #3 and #4 operated effectively throughout the remainder of the 12 month pilot study. Sensors #5 and #6 slowly degraded until total failure after 75 days of operation. As indicated by these results, sensor output can be erratic and inconsistent. Similar examples of sensor problems have been reported in a recent Environmental Technology Verification project (USEPA/NSF 1998) during drinking water studies. In addition, Snicer et al., (1998) compared UV sensor output to actual microbial inactivation, using MS-2 bacteriophage, and found that in any given UV reactor, each sensor will have a unique correlation to actual organism inactivation.

Chemical UV Dosage Determination

The benchmark for all UV irradiance and hence dosage measurements has long been chemical actinometry. A chemical actinometer is a laboratory prepared solution that produces a measurable, stoichiometric response to photons during irradiation. This stoichiometric response can then be used to back calculate the UV irradiance. The most common chemical actinometers used in UV research are potassium ferrioxalate or malachite (Harris, 1987b); uridine (Linden and Darby, 1997) and peroxydisulfate/t-butanol (Hoyer et al., 1992). Chemical actinometry is an excellent tool for radiometer calibration and small bench scale laboratory studies. However, chemical actinometry is impractical for pilot and full-scale applications since actinometric reagents are too costly, too toxic, or too light sensitive to be used during pilot or full-scale testing.

Mathematical UV Dosage Determination

Modeling UV dosage requires careful consideration of the irradiance distribution and the hydraulic distribution within a given UV reactor. The first successful modeling approximation for irradiance distribution was developed in 1970 (Jacob and Dranoff 1970) and then refined and applied to wastewater UV reactor design in the 1980's (Scheible et al. 1986; Suidan and Severin 1986). This model was termed the Multiple Point Source Summation Approximation or MPSS model. More recently, Blatchley and coworkers (1995, 1997) expanded this model by introducing a line source summation that takes the MPSS to the limit of an infinite number of point sources. Bolton (1998) has proposed further modifications to the work of Blatchley (1997) by adding terms to account for reflection and refraction as a beam of UV light passes through the air/quartz/water interface. Combining the sophisticated irradiance models with modern computational fluid dynamic models is a very promising technique for improving UV dosage approximations and enhancing UV reactor designs. Modeling efforts have two major limitations: 1) it is difficult to adequately calibrate and verify sophisticated models due to limitations in UV sensor and in-situ velocity measurement technologies, and 2) the refined models, while hoping to markedly improve reactor designs, cannot provide routine performance monitoring of operating UV reactors.

UV Disinfection Effectiveness

Efforts by Linden (Linden, 2001) and the USEPA UV Guidance Manual preparation team (USEPA, 2003b) to review UV literature and prepare detailed dose (I x T) tables for the USEPA Stage 2 rulemaking have resulted in a detailed literature review, which has been updated and expanded upon here.

Pathogens of concern in a water treatment system consist of bacteria, viruses, and protozoans. According to the available literature UV irradiation is effective for inactivation of all three types of pathogens at practical doses. Doses for inactivation of bacteria up to 4 $\log_{10}$ range from 3 (*Vibrio cholerae*) to 11 (*Klebsiella terrigena*) mJ/cm^2, depending on the species. Doses for inactivation of viruses up to 4 $\log_{10}$ range from 16 mJ/cm^2 (Hepatitus A virus) to 200 mJ/cm^2 (Adenovirus 2). Doses for inactivation of *Cryptosporidium parvum* oocysts up to 4 $\log_{10}$ range from between <10 to <19 mJ/cm^2. Doses for inactivation of *Giardia lamblia* up to 3 $\log_{10}$ range from between 2 to 20 mJ/cm^2. Broadly classified, the relative UV susceptibility of these groups of microorganisms is bacteria>protozoan>viruses.

To put these numbers in perspective, UV systems for water disinfection in Europe are designed for a delivered dose of approximately 40 mJ/cm^2. UV systems for wastewater reclamation to meet California Title 22 standards are designed for a delivered dose of approximately 140 mJ/cm^2. Numerous systems are currently in operation at these doses, therefore some precedence for these system designs exists.

Theory suggests that if UV irradiation can reach the genetic material of a microorganism such that the nucleic acids absorb UV, that microbe will experience some level of UV induced damage that may result in the inability of the nucleic acid to replicate and thus not allow the microorganism to reproduce. An organism that cannot reproduce, will not be able to infect a host, and will eventually die. In the past 15 years, there have been many studies investigating the efficacy of UV for inactivation of a variety of pathogens and indicator organisms. Many organisms of public health significance have been studied by numerous investigators, resulting in a number of independent data sets for inactivation. Although there are instances of some outliers, the majority of the data sets available suggest similar conclusions with regard to the efficacy of UV irradiation for inactivation of a given organism. The data reviewed in this section have been verified by a number of independent researchers and thus hold significant merit.

Introduction

The data reviewed here are for studies using monochromatic low pressure UV irradiation – unless specifically noted. There are a number of advantages to comparing UV efficacy studies that employ low-pressure UV lamps. The systems are relatively simple to set up and use and the lamps are readily commercially available. Most importantly the UV dose for such systems can be measured and reported with confidence as the dose at 254 nm for a given inactivation. Dose measurement is performed by measuring incident UV irradiance at the surface of a complete mixed batch sample with a radiometer and detector calibrated to standards traceable to the National Institute for Standards and Technology (NIST). The incident irradiance is then corrected for water quality induced UV demand to calculate an average UV irradiance. This average irradiance multiplied by the time of exposure is the UV dose delivered.

Most studies were performed in laboratory or finished/tap water with spiked pathogen samples, typical of bench scale work unless otherwise noted. The doses are typically reported as average delivered doses or incident doses. For clean water samples (85 to 95 percent UV 254 transmittance) and at a typical sample depth of 1 cm, the difference between average and incident doses is negligible (less than 9 percent).

Bacteria

The efficacy of UV irradiation for inactivation of bacterial species has been investigated for many pathogenic and non-pathogenic strains. Based on a survey of data available in the literature (Table A-1 of the UV Guidance Manual- Draft) the UV dose in mJ/cm^2, for a 2 log_{10} inactivation, ranges from 1 to 12 and for a 4 log_{10} inactivation ranges from 3 to 34 mJ/cm^2. A comprehensive study of numerous pathogens by Wilson et al. (1992) indicated that for a 4 log_{10} inactivation, *Aeromonas hydrophila*, *Campylobacter jejuni*, *E. coli* 0157H7, *Legionella pneumoniae*, *Salmonella typhi*, *Shigella dysenteriae*, *Vibrio cholerae*, and *Yersinia enterocolitica* required respective doses of 5, 4.6, 5.6, 9.4, 8.2, 3, 2.9, and 4.6 mJ/cm^2. Chang et al. (1985) found that *E. coli*, *Salmonella typhi*, *Shigella sonnei*, and *Streptococcus faecalis* required UV doses of 8.4, 7.1, 8.2, and 11.2 mJ/cm^2, for a 4-log_{10} inactivation. Studies of numerous *Salmonella* species by Tosa and Hirata (1998) (*S. anatum, S. derby, S. enteritidis, S. infantis, S. typhimurium*) indicated doses ranging from 5 to 15 mJ/cm^2, for 3 or 4 log_{10} inactivation. Among the more resistant bacteria reported in the literature, *Citrobacter freundii*, *Enterobacter cloacae*, and *Klebsiella pneumoniae* require doses of 22 to 34 mJ/cm^2, for a 4-log_{10} inactivation. All of these organisms were studied by Zemke et al. (1990), who also reported a 4-log_{10} inactivation dose of 25 mJ/cm^2, for *E. coli* (ATCC strain 11229). This same strain of *E.coli* was reported by three other independent researchers (Harris et al. 1987a; Chang et al. 1985, Sommer et al. 1998) to require a dose between 5 and 8.4 mJ/cm^2, for a 4-log_{10} inactivation. This inconsistency suggests that the actual doses for inactivation of these supposedly more resistant bacteria may be mistakenly high. Following a review of the research, it appears that the exposure and dose measurement techniques were not typical, thus these results remain in question. Data for inactivation of heterotrophic plate count (HPC) bacteria indicate that these bacteria are no more resistant to UV than other bacteria. Furthermore, these organisms are not considered pathogenic. Typical UV doses required for a 3-log_{10} inactivation of indigenous (non-spiked) HPC are on the order of 10 mJ/cm^2 (with pulsed-UV system) with <10 cfu/mL remaining after this dose (Mofidi et al. 1999). More work on HPC is currently scheduled on at least one other research project. Of the pathogens studied, none appear to be significantly resistant to UV irradiation. Therefore the existing data indicate that UV would be an effective control for most bacterial pathogens.

Potential for Reactivation

Because UV irradiation does not immediately result in cell lysis, there is some opportunity for a UV inactivated cell to regain viability through a repair mechanism. One such repair mechanism is termed "photoreactivation" (Jagger 1958). Following UV inactivation, exposure to light in the 350 to 450 nm range (present in sunlight) could trigger an enzyme mediated process through which some bacteria may be able to repair the UV damage inflicted on the DNA and become viable. Not all bacteria possess this "photoreactivating enzyme" therefore not all bacteria can repair the UV damage. Furthermore, the ability for total coliform to photoreactivate is indirectly related to the applied UV dose. Lindenauer and Darby (1994) concluded that an increase in UV dose beyond that necessary for initial inactivation is extremely valuable in minimizing photoreactivation events. As the UV dose applied was increased, the ability to recover decreased linearly up to a dose of approximately 140 where no reactivation was found. Similarly, a short exposure to photoreactivating light is less likely to result in significant repair taking place. Baron (1997) reported that a 1 $\log_{10}$ increase for the fecal streptococci should be designed for in wastewater systems operating at a dose of 40 mJ/cm^2. A similar amount of photoreactivation was reported by Scheible et al. (1986). Most of the data available for photoreactivation have been collected for coliform organisms studied. However, organisms found to photorepair include total and fecal coliform, *E. coli, Streptococcus feacalis, Streptomyces, Saccharomyces, Aerobacter, Micrococcus, Erwinia, Proteus, Penicillium,* and *Nuerospora* (Whitby et al. 1984; U.S. EPA 1986; Harris et al. 1987a). Organisms demonstrating no tendency to photoreactivate include bacteriophage, somatic coliphage, *P. aeruginosa, Clostridium perfringens, Haemophilus influenzae, Diplococcus pneumoniae, Bacillus subtilis*, and *Micrococcus radiodurans* (Harm 1980; Whitby et al. 1984; U.S. EPA 1986; Harris et al. 1987a; Palmateer and Whitby 1987; Chrtek and Popp 1991). If photoreactivation were expected to be a concern, increasing the dose for inactivation may be warranted. However, if systems were designed to meet virus removal, the doses utilized would likely be adequate to eliminate the potential for significant reactivation.

Reactivation of bacteria is not expected to be a problem for treated drinking water when a covered clearwell before the distribution system is utilized. Furthermore, if UV irradiation is followed by a chlorine or chloramine step to provide some residual disinfectant, the chance for a reactivation event to occur is negligible. However, data to support these claims is not yet well developed but is underway by more than one research group.

Viruses

Viruses as a group appear to be the most UV resistant organisms and will likely be the limiting organisms in determining a reactor design dose for systems requiring virus inactivation. Currently, California Title 22 standards for design of UV systems for water reclamation require a dose of 140 mJ/cm^2 based on a 4-$\log_{10}$ inactivation of Poliovirus (including a safety factor). Seven independent research groups (Chang et al. 1985; Harris et al. 1987a; Maier et al. 1995; Meng and Gerba 1996; Snicer et al. 1998; Tree et al. 1997; Wilson et al. 1992) all reported a dose range of 20 to 30 mJ/cm^2 was required for a 4-$\log_{10}$ inactivation of Poliovirus Type 1. Results reported by Tree et al. (1997) were for wastewater with a turbidity of 20 NTU. Battigelli et al. (1993) reported that Coxsackievirus required a dose of 20.6 mJ/cm^2, for a 3-$\log_{10}$ reduction, while Hepatitis A virus required a dose of 16.4 mJ/cm^2, for 4 $\log_{10}$ reduction. Other researchers (Snicer et al. 1998; Wiedenmann et al. 1993; Wilson et al. 1992) reported similar results indicating that the Hepatitis A virus required a UV dose of between 7 and 30 mJ/cm^2 for 4-$\log_{10}$ reduction. Three research groups reported that Rotavirus strain SA 11 requires a dose of 23 to 26 mJ/cm^2 for 4 $\log_{10}$ inactivation (Battigelli et al. 1993; Chang et al. 1985; Wilson et al. 1992). Rotavirus strain WA was reported to require a dose of 50 mJ/cm^2 for 4 $\log_{10}$ reduction. The most resistant virus type appears to be Adenoviruses. Meng and Gerba (1996) have reported that adenovirus serotypes 40 and 41 exhibited significant resistance to UV, requiring doses up to 121 mJ/cm^2, for 4 $\log_{10}$ inactivation. Additional work by Gerba et al. (2001) has shown that Adenovirus serotypes 2, 5 and 7 require doses of up to 200 mJ/cm^2 for a 4 $\log_{10}$ inactivation. These data are currently being corroborated and early results indicate similar dose ranges (Yanko 2000)

These differences in virus susceptibility may be related to the morphology of the virus, such as the type of nucleic acid present (single versus double stranded RNA or DNA) or the lipid/protein envelope. Studies are currently underway to systematically investigate the effects of virus morphology on disinfection effectiveness. As viruses exhibit the most resistance to UV, there are numerous studies underway to augment the existing database and better understand the fundamental basis for the relative resistance of different types of viruses and between viruses and other types of microorganisms.

Cryptosporidium parvum

Previous studies based on viability assays such as excystation and dye inclusion/exclusion reported that UV doses greater than 8,000 mJ/cm^2 were necessary for a 2 to 3 $\log_{10}$ inactivation of *Cryptosporidium parvum* (Campbell et al. 1995), thus UV was not considered a viable technology for disinfection of protozoan pathogens. It was first determined in 1996 that pulsed-UV radiation was effective for inactivation of *Cryptosporidium parvum* when assayed using a mouse infectivity endpoint (Arrowood et al. 1996). However, it was not until 1997 when Clancy and colleagues (Clancy et al. 1998a) reported that pulsed UV was effective for *Cryptosporidium* inactivation, that this information was widespread in the water industry and since then numerous studies have followed to refine the UV dose-inactivation response for this pathogen.

More than six independent data sets now exist for inactivation studies with *Cryptosporidium parvum* using either animal infectivity or cell culture infectivity endpoints. These studies originated in research groups at Clancy Environmental Consultants (mouse infectivity – Clancy et al. 1998b,c,d; Bukhari et al. 1999; Hargy 2000), Center for Disease Control (mouse infectivity - Arrowood et al. 1996), University of South Florida (cell culture - Slifko et al. 1997), University of North Carolina/Duke University (cell culture – Shin et al. 1999, 2000a), University of Alberta (mouse infectivity – Finch and Belosevic 2000), and Metropolitan Water District of Southern California (cell culture – Mofidi et al. 1999). All the data (not including the original pulsed-UV work that only looked at doses >1,000 mJ/cm^2) conservatively suggest that a UV dose of 20 mJ/cm^2 will result in at least a 3-$\log_{10}$ inactivation of *Cryptosporidium parvum*. In fact, in many studies, a dose of 10 mJ/cm^2 resulted in the assay detection limit being reached.

Early pulsed UV work in 1996 (Arrowood et al. 1996) and 1997 (Clancy et al. 1998a,b; Slifko et al. 1997) only investigated very high UV doses (> 1,000 mJ/cm^2) and the dose measurement techniques were not validated substantially. This early work led to varying hypotheses that perhaps the pulsing nature and the extremely high energy of the light, or the polychromatic nature of the pulsed-UV spectra was responsible for the susceptibility of C. parvum. These studies provided momentum for further investigation of *Cryptosporidium. parvum* inactivation with other UV sources. Subsequent pulsed UV work by Clancy et al. (1998b) and Mofidi et al. (1999) illustrated that this technology also worked well at UV doses relevant to water treatment in tap and finished water respectively.

Bukhari et al. (1999) reported that medium pressure UV was effective at inactivating *Cryptosporidium parvum* up to 3.9-log_{10} inactivation at a UV dose of 19 mJ/cm^2 in a 200 gpm pilot system, using a mouse infectivity assay. Hargy (2000) reported > 4.5 log_{10} *Cryptosporidium parvum* inactivation in recycled filter backwash supernatant (turbidity = 16 NTU) at a UV dose of 3 mJ/cm^2 on the bench scale. Hargy also tested recreational water at pilot scale with medium pressure UV at a dose of 10 mJ/cm^2 (40 gpm) and reported greater than 4.4-log_{10} inactivation. Subsequent medium pressure bench work by Finch and Belosevic (2000) illustrated that a germicidal dose of <10 resulted in approximately 2.5 log_{10} inactivation. This medium pressure UV work illustrated that a continuous wave UV source could perform as well as a pulsing source. Thus pulsing of UV was not necessarily the attribute that resulted in UV effectiveness.

UV irradiation from a monochromatic low pressure UV source was subsequently investigated by Shin et al. (1999, 2000a) for inactivation of *Cryptosporidium parvum* and illustrated that a dose as low as 4 mJ/cm^2 resulted in >3-log_{10} inactivation. Hargy (1999) also reported that a low pressure UV dose of 10 mJ/cm^2 on the bench scale resulted in a 4 log_{10} inactivation. Thus monochromatic UV appeared to be as efficient at inactivation of *Cryptosporidium parvum* as medium pressure and pulsed UV sources.

These studies in sum, utilizing low pressure (monochromatic), medium pressure (polychromatic) and pulsed (polychromatic) UV systems on the bench and pilot scale in a variety of water matrices, clearly indicate that UV photons are effective for the inactivation of *Cryptosporidium parvum*. *Cryptosporidium* is perhaps the pathogen most completely studied for UV effectiveness and is currently under further study.

Reactivation of Cryptosporidium

In a controlled study of the potential for *Cryptosporidium parvum* to repair via dark or light mediated pathways, there was no evidence of any reactivation in multiple trials (Shin et al. 2000a). Reactivation conditions tested were optimized for both prokaryotic and mammalian repair and recovery. UV doses used for inactivation prior to reactivation attempts were low, approximately 1 to 2 mJ/cm^2, and resulted in approximately 0.75 to 2 log_{10} inactivation. The low doses and optimum reactivation conditions were chosen to allow the best chance for reactivation to occur. However, no increase in survivors following reactivation protocols was evident.

Giardia

Similar to the story with *Cryptosporidium*, initial studies of the UV inactivation of *Giardia* was done using viability assays (excystation) and reported UV doses were relatively high and considered impractical for use of UV for *Giardia* control. Karanis et al. (1992) Carlson et al. (1985) and Rice and Huff (1981) all reported UV doses necessary for 1 to 2 $\log_{10}$ inactivation at 60 to 120 mJ/cm^2 based on excystation viability assays. Although these doses are not astronomical, they were considered impractical when *Giardia* could be effectively controlled using moderate doses of chlorine.

UV disinfection of *Giardia lamblia* (human pathogen) has been or is currently under investigation by at least four independent research groups including University of North Carolina/Duke University/McGill University, University of New Hampshire/Tufts University, Safe Water Solutions, Metropolitan Water District of Southern California. All groups are utilizing an animal model (mongolian gerbils) for infectivity assays. Although this data is not yet completely out in the peer-reviewed literature, some has been presented at recent conferences and is soon to be published. The data indicate that conservatively, a dose of 20 mJ/cm^2 results in at least 3 $\log_{10}$ inactivation of *Giardia lamblia.* Shin et al. (2000b) reported that in repeated trials, a dose of approximately 4 mJ/cm^2 of low pressure monochromatic UV resulted in >3 $\log_{10}$ reduction in infectivity in lab water. Malley (2000a) reported that light from a pulsed UV source achieved greater than 5 $\log_{10}$ inactivation of *Giardia lamblia* in secondary effluent and biological nutrient removal treated wastewater at a UV dose of approximately 8 mJ/cm^2. A study by Mofidi et al. (2001) indicated that indeed both *Giardia lamblia* and *Giardia muris* are both highly susceptible to low doses of UV radiation. Their results showed that at UV doses greater than 3 mJ/cm^2, greater than 2 $\log_{10}$ (99 percent) inactivation was observed for each organism. More research is necessary however, to investigate assay variability and the effect of cyst clumping on these results.

UV disinfection of *Giardia muris* was or is currently under investigation by at least three independent research groups including University of Alberta, CH2MHill, and Montgomery Watson. All groups are utilizing a mouse infectivity model. This data is not yet completely out in the peer reviewed literature and the results are slightly more varied but still indicate conservatively that a dose of less than 10 mJ/cm^2 corresponds to >2 $\log_{10}$ inactivation for medium pressure polychromatic UV radiation. Finch and Belosevic (2000) reported that a 2.5 $\log_{10}$ inactivation of *Giardia muris* could be achieved at a medium pressure UV dose of approximately 10 mJ/cm^2. However, this group reported a tailing effect (inactivation did not increase above 2.5 $\log_{10}$ with increasing dose) beyond this value. Soroushian (2000) reported that a dose of 10 resulted in >4 $\log_{10}$ inactivation for low and medium pressure UV lamp sources. Oppenheimer (2000) reported initial results indicated a dose of 10 mJ/cm^2 resulted in >2 $\log_{10}$ inactivation.

Photoreactivation of Giardia

Photoreactivation experiments for *Giardia lamblia* cysts conducted by the University of North Carolina/Duke University group (Shin et. al. 2001a) indicated there was no evidence of either light or dark repair of UV-damaged DNA, for experimental UV doses of 16 and 40

mJ/cm^2. This study was conducted under both bacterial and mammalian DNA repair conditions, as such, more investigation is required under a variety of environmental conditions to determine the likelihood of cell DNA repair. Additional study is currently underway by the University of North Carolina/Duke University group. Preliminary data based on trip blanks sent in the dark suggest there is no occurrence of dark repair (Shin et.al. 2001b).

Comparative efficacy of monochromatic and polychromatic UV systems

Much of the pathogen inactivation data reported in the literature is based on inactivation with low-pressure monochromatic UV radiation at 253.7 nm. Medium pressure and pulsed UV lamps systems supply polychromatic UV radiation at wavelengths throughout the 220 to 300 nm germicidal UV range (any beyond). Linden et al. (2000) investigated the relative wavelength effectiveness of *Cryptosporidium parvum* and these data appear to match the historical wavelength effectiveness data in the literature for *E. coli* (Gates, 1929). Both data sets suggest that the most efficacious wavelengths for inactivation are between 260 and 270 nm. Irradiation at 253.7 nm is approximately 90 percent as efficacious as radiation at these peak wavelengths. The effectiveness of UV at wavelengths above 270 nm and below 245 nm is significantly lower. Because medium pressure UV systems emit radiation at other less efficacious wavelengths (such as in the 220 to 240 and 280 to 300 nm range) there is not necessarily an inherent advantage to use of one system over the other for these pathogens. In a side by side bench study of low and medium pressure UV system efficacy for inactivation of *Cryptosporidium*, they both performed similarly at germicidal doses of 2 mJ/cm^2 (Linden et al. 2000).

The relative wavelength effectiveness data generated for bacteriophage MS2 differs from the *E. coli* and *Cryptosporidium* data in that the most effective wavelengths are around the 270 nm range and near 216 nm. It is possible that given the difference in morphology and nucleic acid structure, viruses may be more susceptible to polychromatic UV radiation than to monochromatic UV at 253.7 nm. Studies to determine the relative efficacy of viruses to different UV sources and the potential impacts that morphology plays in relative resistance of microorganisms to UV disinfection are underway.

Impacts of Water Quality on UV Effectiveness

When a UV dose is specified to achieve a level of disinfection, that applied dose can be designed for, in a relative sense, regardless of water quality. Once a UV dose is set for a given system, changes in water quality outside the bounds of what the system was designed for may compromise the effectiveness of the UV system. Modern designs of UV systems allow for some level of irradiance adjustment (power level adjustment) to adapt quickly to changes in water quality and assure delivery of designed dose. Nonetheless water quality can affect, either directly or indirectly, the ability of a UV system to achieve a desired designed-for outcome. Water quality factors that may affect UV disinfection effectiveness are UV absorbance, temperature, pH, dissolved ions, and turbidity/particles,. Direct and indirect effects of these factors are discussed below.

UV absorbance/UV Percent Transmittance

UV light becomes attenuated in liquid as a function of absorbance and distance from the UV source. The effects of UV absorbance on the UV energy available at a given distance from a UV source can be determined using the Beer-Lambert Law. UV photons are not distributed evenly throughout a UV reactor in the same way that a chlorine chemical is mixed and dispersed. UV photons originate from fixed sources (lamps) within a reactor, thus the distribution of photons within a reactor is dependent upon the orientation of the lamps and the size of the reactor. An irradiance distribution field can be mapped for a given reactor configuration and UV absorbance of a water. Typically, the closer to a lamp a target gets, the higher the irradiance (fluence rate) it will experience. The lowest fluence rates are experienced at or near the reactor walls. As UV absorbance increases, the ability of UV photons to penetrate through a liquid is compromised. Thus reactor UV dose is adversely affected by UV absorbance.

Temperature and pH

Typical water quality parameters that impact chemical disinfectants such as temperature or pH do not directly affect delivery of UV photons or compromise the disinfecting power of a UV photon in the same way. In a study on the effect of temperature and pH on disinfection of MS-2 coliphage, Malley (2000b) found that neither temperature (1 to 23°C) nor pH (6-9) affected the inactivation of this viral indicator. Temperature may affect performance of a disinfection system indirectly. Under extremely cold-water temperatures, low-pressure UV lamps that normally run at 40 to 60°C and were initially rated in warmer water exhibit slight decreased power outputs. (Mackey et al. 2000) This effect was not seen for medium pressure UV lamps that tend to run at 400 to 600°C. If low pressure or low-pressure high-output technology is to be used, the design criteria must address the slight reduction in lamp output at the coldest water temperatures anticipated, while evaluating different systems. Effects of pH may also be indirect with respect to the precipitation of ions from water onto lamps causing fouling of the quartz sleeves.

Dissolved ions

Total dissolved solids in the water may affect the fouling of lamp sleeves. If the water has moderate to high hardness, then the presence of calcium, magnesium, or iron will accelerate fouling of lamps and sensors. This problem is well recognized as an operational hurdle for UV systems and solutions have been effectively engineered to deal with fouling by UV manufacturers. The rate of fouling is directly related to temperature of the quartz sleeve at the quartz-water interface. Low pressure UV lamps, due to the relatively low operating temperature, exhibit slow to moderate fouling that can be controlled by manual cleaning with an acid wash or other detergent on a regular basis (on the order of weeks or months) as part of the UV system maintenance. Medium pressure UV lamps operate at very hot temperatures, thus cause precipitation of dissolved ions onto the lamp quartz surface. These systems necessitate mechanical wipers that can clean the fouled sleeves (on the order of hours) in-situ on a regular basis.

Softening

Because hardness is a water quality parameter that is undesirable for UV disinfection, softening should be a welcome improvement for UV disinfection. Softening by an ion exchange process would have no perceived negative impacts on UV disinfection. In fact ion exchange would likely improve performance due to decrease of fouling constituents and potential sequestering of particles in the ion exchange column. Softening by lime-soda ash is a precipitation process. Thus the formation of floc precipitate could have a negative impact on the performance of UV by an added load of particulates if there is poor settling or no filtration present. However, because these systems have to meet the same effluent turbidity criteria for filter performance as conventional plants these negative scenarios should not be an issue. Consequently, the turbidity of water for softening systems would not typically be more than 0.1 to 0.2 NTU higher that that of conventional systems. Mackey et al. (2000) reported finding no differences in performance between pilot UV systems treating softened or unsoftened water.

Unfiltered systems

Although UV disinfection is well proven for filtered water disinfection, research on unfiltered systems is not as developed, although numerous projects are currently underway. Unfiltered systems typically produce water in the 1 to 2 NTU turbidity range, although up to 5 NTU is currently allowed. In this range the particles themselves are unlikely to affect greatly the delivery of UV dose due to scattering or UV absorbance. In one study by Malley (2000a) it was found that the addition of particulates from wastewater or alum floc up to a turbidity of 2, increased the dose necessary for 2 $\log_{10}$ inactivation of MS2 bacteriophage by 50 percent. Increase in turbidity up to 5 NTU increased the dose necessary by approximately 100 percent.

In another study conducted by Malley (2001), using clay turbidity up to 12 NTU and algal content of 42 μg/L as chlorophyll-a, there was little change in UV dose. In a study by Linden and his colleagues (2001), it was found that UV dose was only minimally affected when non particle wastewater was used to create water with turbidity between 1 and 10 NTU with particle size less than 40μm, 11μm, and 5μm. However, during the design of a UV disinfection system, collection of water quality data will allow for systems to be designed to account for particulate interference with UV photons. This design approach would be the same for waters of higher turbidity and has been used for the design of wastewater systems where particulates are a fact of water. If a UV system that is designed around a clean water and the upstream processes are changed such that the water now contains significant amounts of particulates, then the designed for UV dose would be compromised. As reported by Hargy (2000) where *Cryptosporidium* was inactivated to greater than 4 $\log_{10}$ (detection limit) with a UV dose of 3 mJ/cm^2 in recycled filter backwash water (NTU > 11), obtaining a specified $\log_{10}$ inactivation of a pathogen or indicator organism is not going to be an issue for a properly designed UV disinfection system. What may be an issue is the capability of inactivating particle associated pathogens possibly produced in unfiltered systems

Particles and Microbe Clumping

It has been illustrated that non-aggregated microorganisms are easier to disinfect than aggregated ones (Parker and Darby 1995). Therefore, water quality parameters or upstream treatment processes that encourage or facilitate such aggregation or association of pathogens with particles or flocs can protect microbes from UV photons. Research has shown that biological flocs formed in wastewater have extremely high UV absorbances and depending upon the upstream processes, may harbor coliform organisms and shield them from UV light. Aggregation and clumping, have been shown to interfere with chemical disinfection of bacteria (Stewart and Olson 1996). This response may be microbe specific and not a ubiquitous trait. However, the typical "tailing" response exhibited in coliform indicator organisms in wastewater at high UV doses is attributed to particle association (Darby et al. 1995). The ability of typical water treatment processes to induce particle aggregation or association with flocs is not well understood. However, because these water treatment processes such as coagulation are not related to a microbial growth process as is the case in activated sludge wastewater flocs, it is expected that the incidence of particle or floc associated microbes would not be significant. Furthermore, presence of filtration would effectively remove the flocculated particles.

In summary, even with a well designed system to deliver a specified dose for inactivation of a target pathogen, the presence of particles may provide some protection to microbes beyond the delivery of any dose. However as shown in wastewater where biological growth may encourage the formation of microbe associated particles, that faction of microbes is relatively small and would not limit achieving a given $\log_{10}$ reduction of pathogens in the bulk water for a properly designed system. To what extent microbe-particle association in water occurs, research has not yet been developed.

CHAPTER 3
PRELIMINARY BENCH SCALE TESTING

INTRODUCTION

Prior to this research, existing literature related to ozone and UV disinfection suggested the following:

1) Ozone is highly effective at inactivating *Giardia*;

2) Ozone is effective at inactivating *Cryptosporidium* but may have large and costly contact time (CT) requirements at cold temperatures;

3) Ultraviolet irradiation is highly effective at inactivating *Cryptosporidium* and is insensitive to temperature;

4) Ultraviolet irradiation is effective at inactivating *Giardia;*

To demonstrate that the combination of ozone and UV can effectively inactivate *Giardia* and *Cryptosporidium* at cold temperatures, batch ozonation studies as described by Finch and Belosevic (1999), followed by batch bench-scale collimated beam MP-UV studies at 0.5 ^{o}C were conducted on winter water samples collected from Wachusett Reservoir.

The purpose of the bench-scale studies was two-fold. First, it was necessary to perform bench-scale collimated beam studies to validate the UV dose provided in the pilot-scale biodosimetry studies that were conducted as part of this project (refer to Chapter 4). Therefore, bench-scale studies were conducted using batch ozonation and collimated UV equipment on MS2 bacteriophage. Second, proof of disinfection concepts and extrapolation of pilot data for bacteriophage to inactivation of the live human pathogens, *Giardia* and *Cryptosporidium*, requires bench-scale studies (Malley, 1998). Bench-scale studies were conducted as follows:

- *Giardia lamblia* studies were conducted at the University of New Hampshire with animal infectivity studies using gerbils performed in the labs of Dr. Widmer at Tufts University School of Veterinary Medicine.

- *Cryptosporidium parvum* studies were conducted in the labs of Clancy Environmental Consultants (CEC) with mouse infectivity performed by Marilyn Marshall at the University of Arizona. Dr. Clancy of CEC also provided input and expert review of all animal infectivity protocols used in this study.

After initial ozone dose tests were conducted to determine the CT needed to provide 2 and 3 log inactivation of *Cryptosporidium*, an experimental design for the ozone, UV, and ozone/UV tests for both *Cryptosporidium* and *Giardia* was developed. The ozone dose tests were performed approximately one week before the inactivation tests so that sufficient data were available to select ozone doses but water quality would not change significantly. The experimental design is outlined in Table 3.1.

Table 3.1
Experimental Design for Bench Scale Inactivation Studies

Cryptosporidium Variable	Test 1 Control	Test 2 Ozone/UV Synergy	Test 3 2 log Inactivation	Test 4 3 log Inactivation
Ozone dose, mg/L	NA	1.5	1.5	2.1
Contact time, minutes	NA	40	40	40
Target Ozone CT mg*min/L	NA	minimum	24	37
Initial Residual, mg/L#	NA	1.3	1.3	1.8
Final Residual, mg/L#	NA	0.3	0.3	0.6
Need for Quenching	No	Yes	Yes	Yes
UV "Dose", mJ/cm^2	NA	5	NA	NA
Giardia Variable	Test 1 Control	Test 2 Ozone/UV Synergy	Test 3 UV 5 mJ/cm^2	Test 4 UV 10 mJ/cm^2
Ozone dose, mg/L	NA	1.5	NA	NA
Contact Time, minutes	NA	40	NA	NA
Target Ozone CT mg*min/L	NA	minimum	NA	NA
Initial Residual, mg/L#	NA	1.3	NA	NA
Final Residual, mg/L#	NA	0.3	NA	NA
Need for Quenching	No	Yes	No	No
UV "Dose", mJ/cm^2	NA	5	5	10

estimated based on preliminary ozone decay studies conducted in February 2002

The remainder of this chapter includes:

- Background information regarding prior studies conducted by MWRA to determine ozone inactivation of *Cryptosporidium*,
- Methods used in the bench scale and field dosimetry testing,
- Ozone dose test data used to select ozone doses at the given contact time of 40 minutes to evaluate inactivation of *Cryptosporidium* under cold water conditions in Wachusett water,
- Results of the *Cryptosporidium* inactivation studies using ozone and ozone/UV,
- Results of the *Giardia* inactivation studies using UV and Ozone/UV and
- Discussion of the findings of the bench scale work.

Background Ozone Inactivation Data

The Massachusetts Water Resources Authority took part in AWWARF project 90797, *Evaluation of Cryptosporidium Inactivation in Natural Waters*. As part of these tests, MWRA developed a water-specific CT curve for inactivation of *Cryptosporidium*. The MWRA data were used as a case study on 3 °C Wachusetts Reservoir water. Testing was performed with various ozone doses, and a first order decay equation was used to model the ozone residual. In order to calculate the cumulative CT product, the estimated residuals were multiplied by the T_{10}

value of each chamber. The CT product was translated into a log inactivation value for *Cryptosporidium* using Table E-10 in the Guidance Manual for Compliance with the Filtration and Disinfection Requirements for Public Water Systems using Surface Water Sources (USEPA, 1990). The results of this study were referenced in several publications dealing with the design of the MWRA ozone facilities and are presented below.

- 2-log inactivation at 3 C = 18 mg*minutes/L
- 3-log inactivation at 3 C = 27 mg*minutes/L

Temperature Correction

The following equation was used in the same AWWARF study to determine the ozone dose and contactor size required to achieve log inactivation of *Cryptosporidium* oocysts.

$$\frac{CT_1}{CT_2} = 1.1644^{(T_2 - T_1)} \qquad (3.1)$$

where CT_1 and CT_2 are the disinfection CT values required at temperatures T_1 and T_2, respectively, to achieve the same log inactivation of *Cryptosporidium*. Using this approach, the CT values for 2 and 3-log inactivation of *Cryptosporidium* at 1° C would be a 35 percent increase, or:

- 2-log inactivation at 1° C = 24 mg*minutes/L
- 3-log inactivation at 3° C = 37 mg*minutes/L

Demand and Decay Studies

Subsequent to the bench-scale inactivation studies, MWRA commissioned studies to determine ozone demand and decay rates. Data from two sets of cold water ozone demand and decay tests (one from January 2000, one from February 2000) were used to estimate the dosages needed to produce the CT values listed above. These tests utilized the six pilot ozone columns located on a skid in the Wachusett Powerhouse. Ozone was applied to the first of the 10-foot tall columns and allowed to decay through the remaining units. Dissolved ozone measurements were obtained from on-line Rosemount dissolved ozone analyzers and from grab samples collected from sample taps located in the side of each column. Grab samples were analyzed using HACH Accuvac ampules. Data from these tests are shown in Table 3.2.

Table 3.2
Wachusett Ozone CT Data

Date	Water Temperature °C	Ozone Dose mg/L	Initial Residual mg/L	Decay Coefficient l/s	CT at 39 minutes mg*min/L	CT at 58 minutes mg*min/L
Jan 12, 2000	5	0.83	0.18	0.057	2.0	2.1
		1.20	0.37	0.039	5.2	6.0
		1.61	0.62	0.028	10.4	12.6
		2.01	0.82	0.022	15.0	18.7
		2.39	0.99	0.017	19.8	25.7
		2.73	1.16	0.015	24.0	31.
Feb 22, 2000	3.3	1.61	0.63	0.026	11.0	13.2
		2.52	1.19	0.017	23.9	31.0
		3.07	1.45	0.016	29.6	38.7
		3.54	1.67	0.014	35.0	46.4
		3.86	1.77	0.013	37.6	50.1
		4.52	2.01	0.013	42.9	57.

The 39 minute contact data from the previous tests are plotted in Figure 3.1. These data indicate that a dosage of approximately 2.5 mg/L should provide a CT value greater than 24 mg*minutes/L while a dosage of 3.5 mg/L should produce a CT value of nearly 37 mg*minutes/L. As these tests were conducted at temperatures of 3 and 5 °C, testing at a lower temperature of 1 °C should have a slower decay rate and slightly higher CT value for the same contact time.

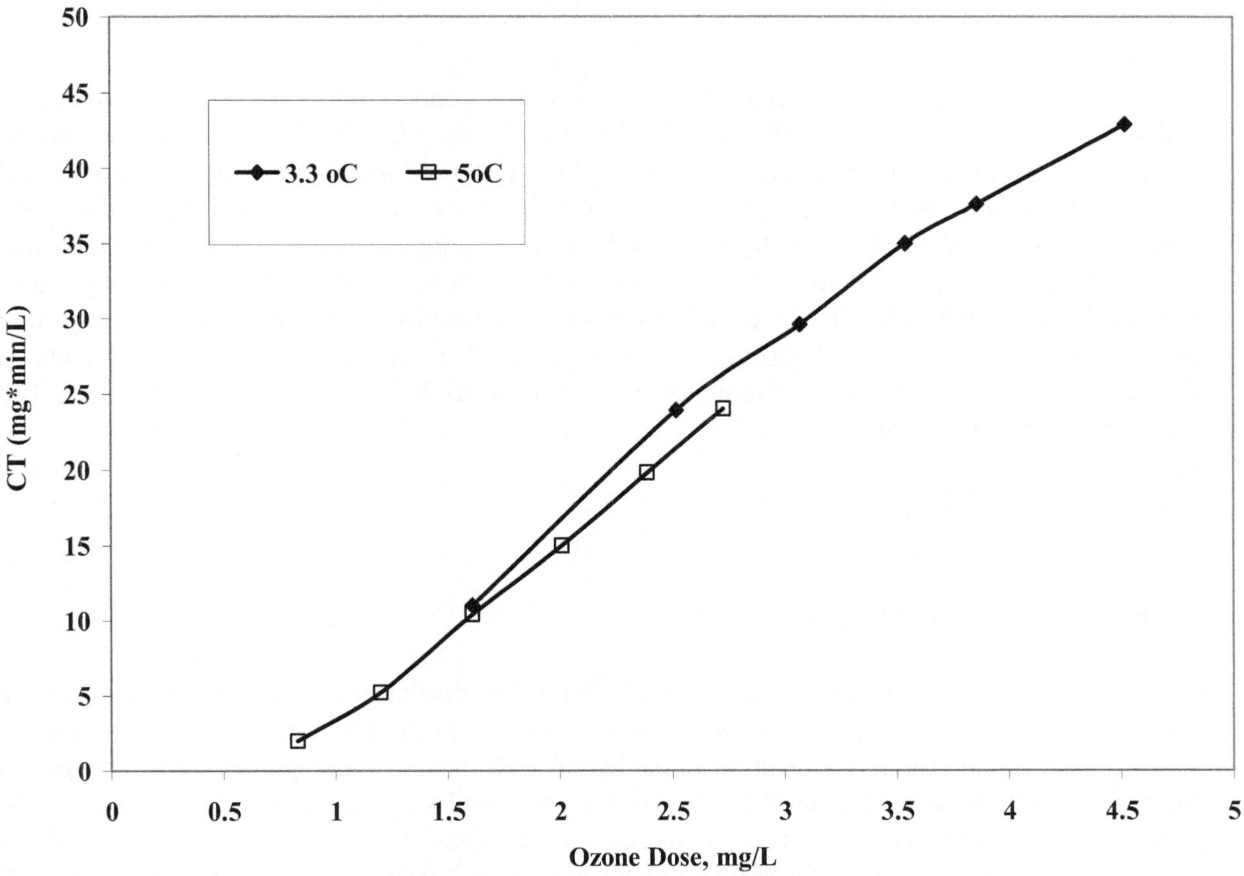

Figure 3.1 CT Values for Prior Testing of Wachusett Water

It was therefore proposed to conduct testing for *Cryptosporidium* inactivation using ozone dosages of 2.5 and 3.5 mg/L. If the decay rate is similar to that found in previous testing, the initial residual for the two tests would range from 1.2 to 1.7 mg/L. After 40 minutes of contact time, the final residual was expected to be between 0.6 and 0.9 mg/L, requiring quenching to eliminate further inactivation.

METHODS

Water Sample Collection and Quality

Samples of Wachusett Reservoir source water were collected by the University of New Hampshire on February 19 and February 26, 2002, and transported in refrigerated glass containers to CEC using overnight express courier. Upon receipt, samples were scanned as described in the following procedure to determine UV absorbance from 220 to 320 nm using a Genesys10 spectrophotometer (Spectronic Unicam, Rochester, NY). As the actual testing would not occur for several days, the purpose of this scan was to provide a baseline measurement that would be compared against a second scan made immediately prior to the disinfection studies.

UV Scans

UV scans are useful to characterize the UV absorbance of the water to be studied. The UV scan conducted on the samples is a modification of the Standard Method 5910B Ultraviolet Absorbance Method where the scan varies over the range of ultraviolet wavelengths from 190 nanometers to 400 nanometers, rather than obtaining the absorbance at 254 nanometers as specified in the analytical method. For the UV scans conducted at the UNH, the absorbance of each sample is measured using matched 1 cm pathlength Suprasil Quartz cuvettes in a double beamed Hitachi 5000 UV-VIS spectrophotometer. The baseline is set using high quality reverse osmosis (RO) water (Type II Laboratory water). Each UV scan was performed in duplicate, with absorbance measured at 1 nm increments from 190 nm to 400 nm. Absorbance measurements can be converted to percent UVT.

$$\%UVT = 10^{-A} \times 100\% \tag{3.2}$$

Ozone Generation and Delivery

Ozone was produced using a CD10/AD corona discharge ozone generator (Clearwater Tech, Inc., San Luis Obispo, CA), which was modified to accept feed gas from a pressurized source. This generator has an output capacity of 1.125 g/hr at 4 standard cubic feet per hour (scfh) with dry gas feed. The feed gas was dry compressed air, regulated to 2 psig, and fed to the generator at 0.5 scfh. The output was delivered to 1 L of test water in an ozone demand-free side arm flask through a glass diffuser inserted through a single-hole stopper. Off-gas was vented from the side arm port and exhausted from the building.

Ozone Residual Measurements

All ozone residual analyses were generated according to Standard Method 4500-O_3 B, (Indigo Colorimetric Method, *Standard Methods for the Examination of Water and Wastewater, 20th ed.*). When dilution of the sample was necessary, ozonated samples were added to a volumetric flask containing de-ionized water (DI) and indigo reagent, and solutions were adjusted to the 100 mL mark with additional DI. Preliminary testing indicated that this procedure resulted in minimal loss of ozone by air stripping during sample manipulation. For residual measurements from test reactions, 10 mL volumetric flasks were used with proportionate reductions of all reagents. All disinfectant concentrations were corrected for dilution factor and expressed as mg/L. To compute the integrated products of disinfectant residual and contact time, ozone residuals recorded over the course of the experiment were multiplied by the contact time interval between each residual measurement. CT values were computed by adding each of these products over the 40 minute course of each experiment.

UV Irradiation Procedure

To assess the effectiveness of UV irradiation for the inactivation of *Cryptosporidium* and *Giardia*, half of the 200 mL oocyst suspension subjected to the lower ozone CT was recovered following quenching and subjected to a UV dose of 5 mJ/cm^2 under batch conditions using a collimated beam format. Based on previously published findings of the effect of UV on *C. parvum* oocysts, this dose was expected to provide at least 3 log additional inactivation (Clancy et al, 2000, 2002; Shin et al, 2001a,b).

A low pressure (254 nm output) UV lamp was enclosed above a solenoid-controlled shutter to regulate light passing through a collimating PVC tube (6.4 cm inner diameter x 30 cm length). The ozone treated oocysts were divided into seven separate aliquots (15 mL each) and placed into 60 mm diameter Petri dishes. Dishes contained 12 mm magnetic stir bars and were placed beneath the collimator tube. Exposure times were controlled by a digital timer connected to the shutter solenoid. Measurements of the UV irradiation reaching the test microorganisms were made with an IL1400A radiometer coupled to an SEL 240 sensor (International Light, Newburyport, MA) traceable to National Institute of Standards and Technology (NIST) standards (254 nm wavelength).

Prior to oocyst exposure, the irradiance distribution across the water surface of the dish was determined by measuring 254 nm irradiance at 0.5 cm intervals along the x-y axis of an 8 cm diameter grid originating at the center of the UV beam and recorded as microwatts per square centimeter ($\mu W/cm^2$). From these readings, the 'petri factor' was determined and used to calculate the average irradiance throughout the entire dish surface using the method of Bolton and Linden (2002) which accounts for surface reflection (2.5%), depth of sample, and UV_{254} absorbance (measured in a subsample of test suspension, post ozone contact and sodium bisulfite addition). Using the average irradiance, the exposure time required for a given UV dose was used to program the digital timer for precise shutter control.

***Cryptosporidium* Oocyst Propagation**

C. parvum oocysts were obtained from the Sterling Parasitology Research Laboratory, University of Arizona - Tucson. The test strain was originally isolated from a calf and is referred to as the Harley Moon, or Iowa strain. The preparation of the oocysts is described in Clancy et al, 2000. The age of all oocysts used in these trials was one month or less. The viability of oocysts was verified upon receipt by *in vitro* excystation assay as described in Robertson et al, 1993. Initial oocyst concentrations were determined by introducing 10 µL samples into a counting chamber (Improved Neubauer hemocytometer, Hausser Scientific, Horsham, PA). A minimum of six chamber counts were performed to enumerate oocysts in the stock solution. Based on these counts, the oocysts were divided into 5 subsets, one for the preliminary demand study, one for each of the two tests, and two controls.

***Cryptosporidium* infectivity analysis**

Following disinfection trials, the samples and process controls were centrifuged at 1050 x g, and oocysts were resuspended to a volume of 1 mL, and transported at 4°C to the animal handling facility at the University of Arizona using overnight express courier. Shipments were accompanied by trip controls containing $1x10^6$ oocysts that had been shipped to and from the laboratory conducting the disinfection trials to identify any changes in oocyst infectivity associated with factors other than disinfection. Upon receipt in the animal handling facility, preliminary hemocytometer counts were performed to determine the concentration of oocysts present in each sample container and to compute the dilutions required to prepare mouse inocula.

Two target inocula levels were selected for each of the ozone and ozone/UV treatments to bracket the anticipated level of inactivation. The oocysts treated with 2.1 mg/L initial ozone dose were delivered to two sets of mice in quantities of $1.0x10^3$ and $1.0x10^5$, to capture inactivation ranging from as little as no inactivation to as much as 4 log. The 1.5 mg/L initial ozone dose oocysts were delivered in quantities $1.1x10^2$ and $1.1x10^4$ per mouse, to measure levels of inactivation ranging from as little as no inactivation to as much as 3 log.

Oocysts treated with both ozone and UV were delivered in quantities of $1.0x10^5$ and $3.0x10^6$ to detect the anticipated level of inactivation that could occur (3 log to 5.5 log). Inocula for the process control and trip control oocysts were 113 and 106 oocysts, respectively, to capture the infectious dose required to infect 50% of the test animals (ID_{50}), based on previous experiences by the research team. Changes in oocyst viability were determined by infectivity assay in 4 to 6 day-old neonatal CD-1 outbred mice as described in Korich et al, 2000. Specimens with parasites were scored positive and those without parasites were scored negative.

***Giardia lamblia* Cysts and Animals**

Purified cysts of *G. lamblia* produced in gerbils were purchased from Waterborne Inc, New Orleans, LA. The purification method described by Stibbs et al. (1988) was applied. Female gerbils were purchased from Charles River, Willmington, Mass. and were approximately 4 weeks of age on delivery. Each group of animals was housed in a separate micro-isolator cage provided with a filter top.

Collimated Beam Work.

The purpose of the collimated beam work was to determine the inactivation or dose-response of the surrogates in the sample water matrix. The surrogate used in this study was MS2 bacteriophage. Results from bench scale collimated beam testing are a critical part of the biodosimetry testing of the pilot units. A collimated beam apparatus developed by Calgon Carbon (Bukhari et al., 1999) in which either a medium pressure or low pressure UV lamp can be mounted was used for the bench-scale phase of the study.

Microorganisms were suspended in 20 mL of the sample water matrix in a polypropylene or other low-microbial absorbing material petri dish and exposed to a series of increasing doses to achieve a range of inactivation from 0 to approximately 6 logs (a minimum of 6 dose steps plus 0). The samples were continuously mixed using a micro stir bar and magnetic stirrer. Work performed at UNH (Malley and Bellester, 2004.) has shown that sample volume and mixing are critical parameters to control in collimated beam work. The adequacy of mixing in the 20 mL sample volume were verified by replacing the petri dish sample with actinometer solution and performing replicate analyses of sub-samples to ensure incident photons are well distributed to the mixed sample.

Results from the current AwwaRF-sponsored project, "Disinfection Efficiency and Dose Measurement for Medium Pressure and Pulsed UV Disinfection Systems," (Linden and Mofidi, Co-Principal Investigators) were used to guide dose determination methods for the MP-UV collimated beam studies. The incident UV irradiance was measured at the center of the liquid surface of the sample. By making appropriate measurements at positions away from the center, the average incident UV irradiance across the top of the petri dish was obtained.

Changing the exposure time varied the UV dose delivered. The average UV dose in the 20-mL sample was calculated using the integration of the Beer-Lambert Law over the depth of the liquid sample as described in Morowitz (1950), adapted for use in suspensions receiving polychromatic radiation (Linden and Darby, 1997). Sample water was used as the water matrix for the collimated beam studies. The surrogate was added to the sample water at a high titer. This was to ensure that some of the surrogate is still viable at the highest delivered dose.

Given the lamp irradiance, a spreadsheet created by Jim Bolton of Bolton Photosciences, Inc. was used to calculate the time needed to achieve average doses of 0, 20, 40, 60, 80 and 100 mJ/cm^2. When calculating the time needed, the spreadsheet accounts for the distribution of light across the water surface, the absorbance of the water matrix, the germicidal effectiveness of 253.7nm, the reflection of light off the water surface, and the irradiance output of the lamp. Enumeration of MS2 was accomplished with the single agar overlay plaque assay (see specific protocol for MS2). After 12-16 hours, the number of plaques present on each plate was counted. Plates having fewer than 30 or greater than 300 plaques were reported as too few to count (TFTC) and too numerous to count (TNTC), respectively.

MS2 Bacteriophage Propagation

A liquid culture of MS2 (approx. 1×10^{10}-1×10^{11} MS2/500 ml T-soy broth) is added to the T-soy broth culture of *E. coli* and incubated at 37°C, no shaking, for 12-18 hours. The MS2 strain used at UNH is from ATCC (American Type Culture Collection), catalog # 15597-B1. Transfer culture to appropriate size centrifuge tube, invert three times and spin @ 10,000 rpm, 4°C, 0 break for 10 minutes. Remove tubes and decant off supernatant, being careful not to disturb the pellet. Save the supernatant. Store the MS2 stock at 10°C. The titer should reduce by a log every 8-12 months.

MS2 Enumeration

Enumeration of the viable MS2 was performed by serially diluting the sample in phosphate buffered saline to the desired dilution point. Diluted samples were dissolved in *E. coli*/ Tryptic Soy Agar (T-soy agar) and incubated at 37°C. Following incubation, MS2 plaques appear as circular clearings in the *E.coli* lawn. Enumeration of the plaques was used to determine the concentration of MS2 in each sample in accordance with accepted microbiological data analysis.

Negative and positive controls are necessary in this assay. The negative control includes the addition of only the *E. coli* to the overlay (no sample) to determine that *E. coli* is growing properly, and if any phage or bacterial contamination is present. An additional control that can also be used to determine these factors can be performed by placing a small volume of the *E. coli* host (no MS2 or overlay) on a T-soy plate and examining the resulting colony morphology. The positive control includes the addition of only the *E. coli* to the overlay (no sample) and subsequent plating. Once the overlay is evenly distributed across the surface of the plate, a small volume of MS2 stock is placed on various spots throughout the surface of the overlay. After incubation, the presence of plaques in these spots demonstrates that the *E. coli* host can effectively be infected by the MS2 phage.

RESULTS

Ozone Dose Testing

Immediately upon receipt and prior to commencing with the ozone reactor tests, CEC scanned the UV absorbance of the test water over the range of 220 to 320 nm. The values obtained on receipt of the sample on February 20, and those measured on February 26 are compared in Figure 3.2. The absorbance of the test water did not change appreciably over the holding period.

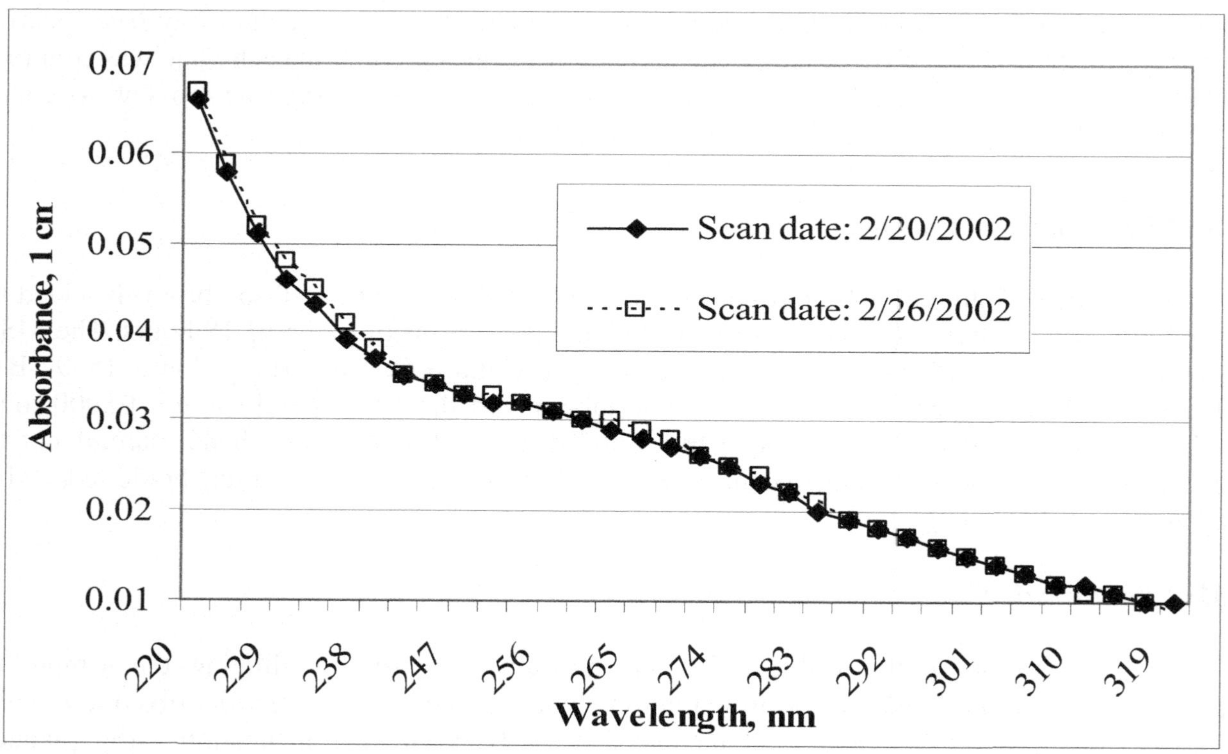

Figure 3.2 Absorbance scans of Wachusett Water, 2/19/2002 and 2/26/2002

Ozone demand was characterized prior to disinfection trials to verify the demand profiles generated in previous studies of this water (January and February, 2000). Concentrated ozone stocks were prepared in Wachusett water at 1°C in the manner described in the prior Methods section, and by transferring sufficient volumes of the concentrated stock to 100 mL batch reactors containing 1°C MWRA Wachusett water to achieve the desired ozone dose. The stock concentration, transfer volumes, and target doses applied are presented in Table 3.3.

Table 3.3
Evaluation of Ozone Decay in Wachusett Water
February, 2002

Target Dose, mg/L	3.5	3.5	2.5	1.0
Temperature, °C	1	1	1	1
Stock O_3 Residual, mg/L	14.9	15.0	15.2	15.9
Reactor Volume, mL	100	100	100	100
mL of Stock to Reactor	23.5	23.3	16.4	6.4
Time (min)		Ozone residual, mg/L		
0.1	3.2		2.1	0.9
0.3		2.9		
0.5				0.7
0.6			2.1	
1.0	3.2	2.8	2	0.6
2.0				0.6
3.0	3.1		1.9	
4.0		2.6		0.7
5.0	2.9		1.8	0.6
6.0		2.2		
10		2.2	1.7	0.5
15				0.4
20				0.5
24			1.1	
30			0.8	0.1
40	1.6	1.1	0.8	

Cryptosporidium Inactivation

The preliminary ozone demand characterizations were used to select initial ozone concentrations and contact times for the *Cryptosporidium* inactivation studies as outlined in the experimental design. The following experimental conditions were selected:

(1) a 2.5 mg/L initial dose with a 40 min contact time to produce a CT of 43 mg·min/L, corresponding to an anticipated 3 log inactivation

(2) a 1.5 mg/L initial dose with a 40 min contact time to produce a CT of 28 mg·min/L, corresponding to an anticipated 2 log inactivation.

Condition 1 (43 mg·min/L) included a 250 mL ozone demand-free glass beaker containing $1x10^8$ oocysts suspended in 80 mL Wachusett water at 1°C. Condition 2 (28 mg·min/L) included a separate ozone demand-free glass beaker containing $2x10^8$ oocysts in 180 mL Wachusett water at 1°C. Following addition of concentrated ozone, test solutions were supplemented as necessary with sufficient de-ionized water to achieve a finished volume of 200 mL. A third reactor was included as a process control. This beaker contained $1x10^7$ oocysts in 100 mL Wachusett water at 1°C.

The process control was subjected to the same manipulations as the test reactors with the exception of disinfectant addition. All reactors were maintained at 1°C in an ice bath with gentle stirring using ozone demand-free magnetic stir bars. After the concentrated ozone stock was added to each reactor and reaction volumes were normalized, the initial ozone residual was immediately determined. Ozone residuals were monitored throughout the test. After a 40 minute contact time, ozone reactivity was quenched by adding 100 µL of sodium bisulfite (3% wt/vol) per 100 mL reaction volume. The effectiveness of the quenching reaction was confirmed by measuring 0 mg/L ozone residual in each test reaction following addition of bisulfite. The quenching agent was also added to the process control. Table 3.4 presents the ozone decay data over the course of the *Cryptosporidium* inactivation tests.

Table 3.4
Ozone Decay during Inactivation Tests

Target Dose, mg/L	2.1		1.5	
Target CT, mg*min/L	37		24	
Temperature, °C	1.0		1.0	
Stock O_3 residual, mg/L	17.6		18.4	
Reactor Volume, mL	100		100	
mL of Stock to Reactor	11.9		16.3	
Time, minutes	O_3	CT	O_3	CT
0.1	1.9	0.2	1.1	0.1
0.5	1.8	0.7	1.1	0.5
1	1.7	1	1.1	0.7
3	1.5	3	0.9	1.8
5	1.5	3	0.9	1.8
10	1.4	7	0.8	4
20	1.1	11	0.7	7
30	0.9	9	0.6	6
40	0.8	8	0.6	6
Integrated CT (mg*min/L)		43		28

The UV absorbance at 254 nm was measured in a sub-sample of the suspension of oocysts in the ozonated test water, following the addition of sodium bisulfite to quench the ozone. The absorbance was found to be 0.060 in the 1 cm path length. The incident UV_{254} irradiance measured at the center of the sample was 80 µW/cm^2. The Petri factor was 0.98. Incorporating these factors, as well as surface reflection (2.5%), an exposure time of 67.2 seconds was calculated to provide a 5 mJ/cm^2 dose.

The number of infectious oocysts delivered to the mice was determined according to the logit dose response model. This method applies a logarithmic transformation that converts the normal dose response curve into a form that can be analyzed by linear regression. Linear regression analysis yields an equation for the straight line of the type

$$y = b + mx \tag{3.3}$$

where b and m are the intercept and slope of the line, respectively. The transformation is accomplished by first defining the term "response logit" for a given oocyst dose as the natural logarithm (ln) of the proportion of mice infected divided by one minus the proportion of mice. That is, response logit = ln[P/(1-P)], where P is the proportion of mice infected with a given dose of oocysts (number of mice infected/number of mice inoculated). The response logit values obtained experimentally were treated as the dependent (Y) variable for regression analysis with the log10 of the number of oocysts in each dose as the independent (X) variable. The Microsoft Excel regression analysis tool was used to perform the least squares regression, provide the regression equation parameters (b,m), and to test the validity of the resulting regression model equation (Korich et al, 2000). The logit dose response model produces a linear regression of the dose response function where the response lies between zero and 100%. Logarithmic transformations of zero and 100% responses are undefined and are therefore not used in the logit model. The linear regression equation of the model,

$$\text{response logit} = -5.62 + 2.81 \log_{10} \text{dose}, \tag{3.4}$$

is a mathematical representation of the dose response function. Here, response logit (y) is calculated from the expression ln(P/1-P) where P is the proportion of animals infected. The intercept (b, -5.62) and the slope (m, 2.81) are provided by linear regression analysis whereas the dose of fresh oocysts (x) is a known quantity. The number of infectious organisms in a given dose of oocysts can thereby be determined when the proportion of infected animals is known. The log inactivation is calculated as log of the number of oocysts in the inoculum minus the log of the number of infectious organisms. Where no mice are infected, log inactivation is reported as greater than the value that would be calculated if a single mouse had been infected. These results are given in Table 3.5.

Table 3.5
Infectivity and $\log_{10}$ inactivation results for *Cryptosporidium*

Test	2.1 mg/L O_3		1.5 mg/L O_3		1.5 mg/L O_3+ 5 mJ/cm^2	
CEC lab ID	20257-2		20257-3		20257-4	
Cyst dose	$1.1x10^3$	$1x10^5$	$1.1x10^2$	$1.1x10^4$	$1x10^5$	$3.3x10^6$
Total Mice	19	20	20	20	19	19
Infected Mice	16	20	5	20	<1	1
% Infected	84	100	25	100	<5	5
Logit	1.67	NA	-1.10	NA	-2.89	-2.89
Oocysts[@] per Dose	395	NA	41	NA	<9	9
Log_{10} Inactivation	0.43	NA	0.43	NA	>4.04	5.55

[@]Infectious

Of the 19 mice receiving $1x10^3$ oocysts treated with 2.1 mg/L ozone, 16 were infected. Application of the logit response model produced an estimated 395 infectious oocysts, corresponding to a 0.43 log level of inactivation. All mice inoculated with $1x10^5$ oocysts treated at this level were infected. All of the mice receiving $1x10^4$ oocysts, and 5 of 20 mice receiving $1x10^2$ oocysts treated with 1.5 mg/L ozone were infected. The logit response model indicated an infectious oocyst number of 41, corresponding to 0.43 log inactivation. The oocysts that were exposed to both ozone and UV (5 mJ/cm^2) caused no infection in any of 19 mice challenged with $1x10^5$ oocysts, demonstrating a minimum of 4.0 log inactivation. One of the nineteen mice receiving an inoculum of $3x10^6$ oocysts exposed to both ozone and UV developed infection, providing a measurable, rather than an inferred, level of inactivation with a more precise measure of disinfection (5.5 log inactivation).

Data from the trip control verified the initial oocyst stock infectivity. Eight of 20 mice receiving trip control oocysts were infected, as was predicted for untreated oocysts (Korich et al, 2000; Clancy et al, 2000, 2002; Bukhari et al, 2000). A comparison of the trip control results with those of the process control confirmed the infectivity of the oocysts. Eighteen of 20 mice receiving process control oocysts were infected, demonstrating that spiking the oocysts into Wachusett test water, subjecting them to laboratory manipulations (vortexing, stirring, shipping and the subsequent analytical processes) had no deleterious effect on their infectivity.

Giardia Inactivation

Three inactivation treatments were performed:

- 5 mJ/cm^2 UV,
- 10 mJ/cm^2 UV,
- 1.5 mg/L dose of ozone with 40 minutes of contact time followed by a UV dose of 10 mJ/cm^2. Control cysts were subjected to each treatment as the experimental cysts, except for the actual exposure and are referred to as process control. The three treatments were performed on

subsequent weeks. Cysts were inactivated and transferred to Tufts on the same day. Upon receipt, cyst solutions were concentrated to 1 ml by low-speed centrifugation. To determine the concentration of the treated and process control cysts, 6 replicate counts were perform using a hemocytometer and the mean cyst concentration calculated. Gerbils were infected orally by delivering a specific number of cysts in aqueous suspension using a pipet. A group of 8 gerbils was infected with 1000 process control cysts each, and two gerbils were mock-infected with sterile water. The remaining 50 gerbils were infected with exponentially increasing doses of treated cysts; 10, 56, 316, 1780, and 10,000 cysts (except for the 5 mJ/cm^2 experiments where 1300 and 5128 cysts were used for the two highest doses).

To identify infections in the gerbils, fecal pellets were collected from each animal on day 6, 8, and 10 post-infection . To ensure that a delayed infection caused by a small number of cysts would not remain undetected, animals that received the highest cyst dose for each treatment were also sampled on day 12 post infection. Fecal smears were stained with a FITC-labeled antibody specific for *Giardia* (Giardi-a-Glo, Waterborne Inc.) and labeled cysts enumerated under an epifluorescent microscope at 200 x magnification.

In addition to infectivity testing in gerbils, cysts were stained with propidium iodide to assess the effect of 10 mJ/cm^2 UV and ozone/UV on the integrity of the cysts (Sauch et al., 1991). Treated and control cysts were suspended in PBS containing 3 μg/ml propidium iodide per 10^6 cysts. Cysts which had taken up the dye were identified by epifluorescent microscopy under UV illumination.

The results from the gerbil infections are summarized in Table 3.6. None of the gerbils infected with treated cysts excreted cysts in the feces. In contrast, all animals infected with 1000 process control cysts developed an infection and produced cysts. All negative control (uninfected) animals, as well as all pre-infection fecal samples were negative. The *G. lamblia* CH3 isolate infectious dose for 50% infectivity (ID_{50}) in gerbils was estimated to be less than 10 cysts (April et al., submitted for publication). The absence of infections in the present study in animals which were administered 10,000 treated cysts indicates that UV and ozone/UV achieved a minimum 1000-fold (3-log) inactivation. Propidium iodide uptake in treated and untreated cysts was identical. In the cyst samples irradiated with 10mJ/cm^2, 28% of the treated and untreated cysts were positive for propidium iodide uptake, and in the combined treatment 21% of the treated and process controls took up the dye, demonstrating that propidium iodide is not a reliable indicator of infectivity for cysts inactivated by UV.

Table 3.6
Infectivity and log_{10} inactivation results for *Giardia*

Test 5 mJ/cm^2	Day 6	Day 8	Day 10	Day 12
5128 cysts/gerbil	0/10$	1/10	0/10	0/10
1300 cysts/gerbil	0/10	0/10	0/10	
316 cysts/gerbil	0/10	0/10	0/10	
56 cysts/gerbil	0/10	0/10	0/10	
10 cysts/gerbil	0/10	0/10	0/10	
10 mJ/cm^2				
10,000 cysts/gerbil	0/10	0/10	0/10	0/10
1780 cysts/gerbil	0/10	0/10	0/10	
316 cysts/gerbil	0/10	0/10	0/10	
56 cysts/gerbil	0/10	0/10	0/10	
10 cysts/gerbil	0/10	0/10	0/10	
1.5 mg/L ozone and 10 mJ/cm2				
10,000 cysts/gerbil	0/10	0/10	0/10	0/10
1780 cysts/gerbil	0/10	0/10	0/10	
316 cysts/gerbil	0/10	0/10	0/10	
56 cysts/gerbil	0/10	0/10	0/10	
10 cysts/gerbil	0/10	0/10	0/10	
Negative Controls	0/2	0/2	0/2	
Process Controls				
1000 cysts/gerbil	8/8			
Trip Control				
1000 cysts/gerbil	8/8			

$ number of infected animals/number inoculated with cysts

DISCUSSION

The anticipated levels of 2 and 3 log oocyst inactivation following ozone treatment at 1°C were not achieved at the CTs used in this study. The measured inactivation with both the low CT value of 24 mg*min/L and high CT value of 37 mg*min/L was less than 0.5 log. These results are similar to the findings in an earlier study of Wachusett water conducted as part of an AWWARF study of *Cryptosporidium* inactivation in natural waters (Oppenheimer et al., 2000). Since these data pertain to the test water studied here and are available for review, it is useful to discuss the results of the present study in their context. In the preliminary report to the MWRA on the specific findings of the 2000 study, a range of inactivation levels from 0.3 to 2.8 log was observed at comparatively high CT values (Oppenheimer, 1997).

The Oppenheimer study examined the effects of ozone on oocyst inactivation at a temperature (3°C) slightly warmer than that studied here (1°C), followed by chloramination to see if the sequential addition of chloramines following ozone treatment would result in enhanced, or synergistic, disinfection as reported by Finch et al. (1997) in laboratory studies using phosphate buffered water. The addition of chloramines provided no additional reproducible disinfection. The inherent variability of animal assays with up to 1 log difference in replicate trials reported by other investigators (Korich et al, 2000; Bukhari et al, 2000) is noted in the earlier MWRA study. The inactivation provided by ozone alone at 3°C ranged from 0.5 to 2.3 log with ozone residuals from 0.9 to 3.4 mg/L. In one trial, a residual of 1.4 mg/L resulted in a 0.5 log inactivation, while in another trial, a residual of 1.2 mg/L ozone led to 1.4 log inactivation. The inactivation after both ozone and chloramine addition, at higher ozone residuals (3.1 to 4.9 mg/L) with a 30 minute hydraulic retention time was also studied. Oocyst inactivation ranged from 0.3 to 2.8 log (average 1.7 log), and was also marked by high variability between replicates. The highest ozone residual, 4.9 mg/L resulted in only 1 log inactivation at 3°C. Even at 18°C, where rapid inactivation kinetics would be expected, with ozone residuals ranging from 1.5 to 3 mg/L, inactivation ranged from 1.6 to 3.4 log, averaging 2.5 log. In contrast to Finch et al (1997), the Oppenheimer study found chloramines offered no additional benefit when applied following ozone in natural waters. Oppenheimer observed that for Wachusett water, "achievement of 3.0-logs of inactivation of *Cryptosporidium* at 3°C appears to require ozone doses and contact times that might not be practically applied".

The inability to achieve 2 or 3 log inactivation using CT values of 24-37 mg·min/L in the current study is consistent with the results of the previous study of the Wachusett reservoir water. The current study temperature was 1°C and the targeted ozone residuals were relatively low (1.5 to 2.1 mg/L). The 0.43 log inactivation observed falls within the expected range based on the historical data.

The additional 5 log inactivation achieved by applying a UV dose of 5 mJ/cm^2 to *Cryptosporidium* is consistent with results of controlled low pressure UV experiments found in the recent literature. Shin et al (2001a,b) reported roughly 1 log inactivation per mJ/cm^2 applied, with 0.8 log at 1 mJ/cm^2, 1.7 log at 2 mJ/cm^2, and >2.6 log at 3 mJ/cm^2. Clancy et al. (2002) investigated five strains of *C. parvum* and the findings also suggest 1 mJ/cm^2 is required for each log inactivation to about 4 log (Table 3.7). In addition, the UV dose for 5 log inactivation ranged from approximately 5 to 10 mJ/cm^2. The results of the present study, which used the Iowa strain, fit well with the known UV inactivation levels, which suggest that the level of inactivation contributed by the ozone under the test conditions studied contributed little to the overall levels of UV inactivation observed here, as these levels can be achieved with UV alone. The verification of ultraviolet irradiation to inactivate *Giardia lamblia* at doses of 5 to 10 mJ/cm^2 indicates that *Cryptosporidium* is not the only protozoan treated effectively with UV.

Table 3.7
Low pressure UV inactivation of *Cryptosporidium* oocysts

C. parvum strain	UV dose (mJ/cm^2)	Log_{10} Inactivation
Iowa	2	3.2
Iowa	4	4.1
TAMU	10	5.2
TAMU	40	>5.5
Maine	5	4.3
Maine	20	5.9
Moredun	10	>5.6
Glascow	5	5.6
Glascow	20	>5.7

Source: Clancy et al, 2002.

CHAPTER 4
FIELD WORK

MWRA FIELD TRIAL

Normal operation of Trains 1 and 2, the two trains involved in the AWWARF Ozone/UV research, involved parallel treatment of Wachusett raw water with UV irradiation for Train 1 and ozone application followed by UV irradiation for Train 2. The water from both primary disinfection schemes then received secondary disinfection with chlorine followed by addition of ammonia for chloramine formation, pH and alkalinity adjustment, and introduction to model distribution systems for comparison of the long term effects of the two primary disinfection processes on the quality of the distributed water. Figure 4.1 presents a schematic of the pilot plant trains, treatment application points and sample locations.

Construction and Operation of the MWRA Pipe Loop Facility

Raw water was withdrawn from the new Cosgrove Intake facility at the lower intake chamber and pumped to the Pipe Loop Facility. The distance from the raw water pump at the intake to the Pipe Loop Facility was approximately 500 feet. The pump is capable of providing 405 gallons per minute (gpm) at 92 feet total dynamic head. The raw water flow could be diverted to a drain via a manual valve, if necessary, or split between the two treatment trains. A raw water sampling point was provided just upstream of the piping split to each train. The flows to Trains 1 and 2 were intended to be constant and continuous at 200 gpm throughout the experimental phase. The raw water flow rate to each train was measured by magnetic meters, with each magnetic meter sending a 4-20 milli-Amp signal proportional to flow to the telog unit which recorded this data.

Ozone was applied to the Train 2 raw water. The ozone feed was a side-stream injection type, i.e., a 30 gallon per minute portion of the Train 2 flow was diverted and pumped through the ozone application system, and then injected back into the remaining Train 2 flow stream.

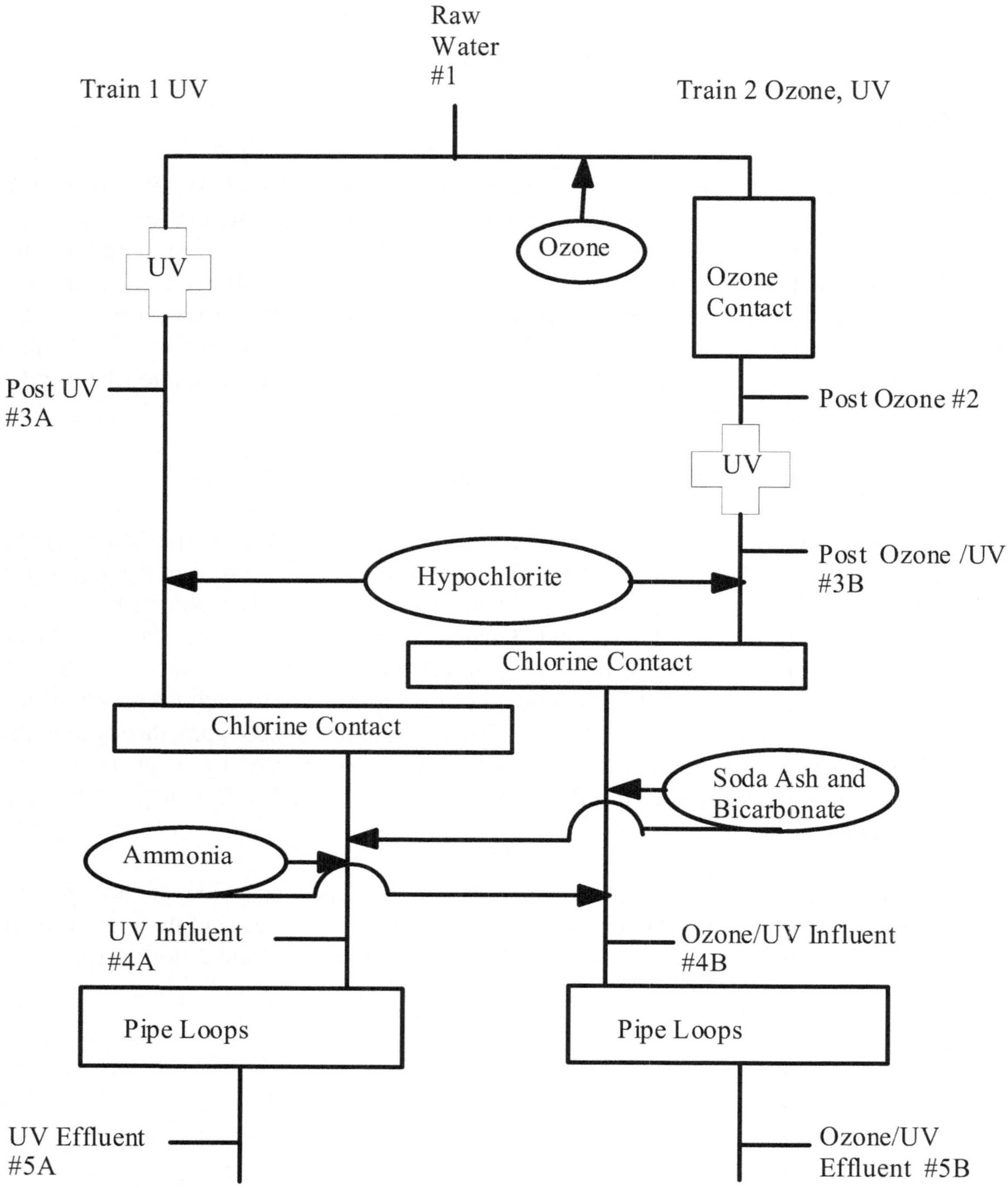

Figure 4.1 MWRA Wachusett Ozone/UV Pilot

An inline static mixer downstream of the ozonated flow injection point ensures that the ozone and supersaturated dissolved oxygen are distributed throughout the 200 gpm Train 2 flow. The ozonated water flows to a set of ozone contact columns that provide up to 30 minutes of contact time at the design flow. The ozone contactor consisted of approximately 120 feet of 36-inch PVC pipe configured such that the operators could manually change the ozone contact time from 10 minutes to 30 minutes. The target ozone dose was 2 mg/L, with a dosage range capability of 1 to 3 mg/L. The ozone generator was capable of producing 8 pounds of ozone per day, which corresponds to a maximum dose of 3.34 mg/L at 200 gpm. Calcium or sodium thiosulfate could be added between the ozone injection point and the UV system, if required, to quench the ozone residual prior to UV treatment. The maximum sodium thiosulfate dosage capability was 2 mg/L

UV irradiation was applied to both Train 1 and Train 2. For Train 1, it was the only form of primary disinfection. For Train 2, UV irradiation followed ozone application. A static mixer with two injection ports for addition of non-pathogenic microorganisms (MS2 bacteriophage) upstream of the UV units was provided to evaluate the effectiveness of the UV disinfection process at different times during the experimental phase. Valving to permit isolation of the UV system was installed to facilitate periodic maintenance or emergency operation. Sample taps were installed before and after each UV unit for collection of water quality samples. The UV units themselves were provided with air release valves to purge any collected gasses during start up and operation. Each UV unit consisted of a reaction chamber containing lamps, sensors to monitor lamp output intensity, and an automated cleaning system. The Trojan Swift 12 UV pilot units had two lamps and two sensors within a flanged tee shaped illumination chamber. The Swift 12 is a medium pressure UV unit designed to deliver a dose of 40 mJ/cm^2. The UV unit required 208 Volt, 3 phase, 60-hertz power to operate the ballasts.

Following UV disinfection, approximately 199.5 gpm of the 200 gpm flow from each train was diverted to waste through an eight inch drain back to the Reservoir.. Chlorine was added to the remaining 0.5 gpm flow from each train. Following chlorine addition, the treated water entered chlorine contact columns; two 12 inch diameter Schedule 80 PVC contact columns were provided, though only one was used for the experiments as one contact column provided 190 minutes (3.17 hours) of contact time before chloramination.

The pH and alkalinity of each flow stream was then adjusted, and the free chlorine residual was converted to the chloramine form by adding ammonium hydroxide before the treated water flowed to the model distribution systems. Each train has two associated pipe loops. Each pipe loop is composed of lengths of 6 inch diameter unlined cast iron pipe, capped at each end with a polyethylene reducer and joined with sections of tubing. There are valves and sample taps at three locations within each pipe rack, and each rack has a flow meter and pressure gauge on the inlet line. Maximum flow through each pipe loop was 1 gpm, and average flow was 0.5 gpm. The average detention time in the pipe loops for each train was 220 minutes or 3.67 hours. At the entrance and end of the pipe loops are taps to send water to annular reactors set up to evaluate growth of attached bacteria. Major sample locations are shown in Figure 4.1 and outlined below:

- #1 Raw water
- #2 Post ozone for Train 2
- #3A Post UV for Train 1
- #3B Post ozone/UV for Train 2
- #4A Influent to the Train 1 pipe loops after secondary disinfection and pH/alkalinity adjustment
- #4B Influent to the Train 2 pipe loops after secondary disinfection and pH/alkalinity adjustment
- #5A Effluent of the Train 1 pipe loops
- #5B Effluent of the Train 2 pipe loops

Table 4.1 presents the operation targets for ozone dose, UV dose, flows, chlorine and ammonia doses, and pH and alkalinity values for the duration of the AwwaRF Ozone/UV pilot study. The ozone dose was manually changed periodically after evaluation of ozone demand. The rationale was that compared with ozone, UV disinfection is more cost effective for inactivation of *Giardia* cysts and *Cryptosporidium* oocysts.

Table 4.1
Target pilot plant operating conditions

Treatment	Variable	Target operating conditions		
		Minimum	Average	Maximum
Ozone	Dosage (mg/L)	1	2	3
	Flow (gpm)		200	
	Contact time (minutes)	7	21	30
UV	Dose (mJ/cm^2)		40	
	Flow (gpm)		200	
	Contact time (seconds)		3	
Corrosion control	pH*	9.3	9.5	9.7
	Alkalinity† (mg/L as $CaCO_3$)		30	
Chlorine/Ammonia	Chlorine residual‡ (mg/L)	1.5	1.75	2.0
	Contact time (minutes)		120	
	Flow (gpm per train)		0.5	
	Cl_2:NH_3 ratio§	4:1	4.5:1	5:1
	Ammonia dose (mg/L)	0.37	0.38	0.40
Pipe loop	Flow (gpm per train)		0.5	

* Na_2CO_3 addition
† $NaHCO_3$ addition
‡NaOCl addition
§ NH_4OH addition

Ozone is used for taste and odor reduction, color removal and UV transmittance enhancement. During this testing, the ozone residual was measured at several points within the ozone contactors to determine CT (for disinfection purposes) along with absorbance to determine the impact of ozonation on downstream UV treatment. The low dose of 0.5 mg/l represents the lower dose limit that might be necessary for oxidation of iron or manganese or for destruction of color or taste and odor causing compounds. The high dose of 3.0 mg/L might be used if inactivation of *Giardia* cysts is desired, especially under warm water conditions when ozone residual decay rates may be rapid. Total chlorine residuals and pH and alkalinity targets for the treated water were selected based on current operating targets for the MWRA system.

The bench scale data from this project and from other sources show that for each mJ/cm^2 UV dose, 1 log of *Cryptosporidium* inactivation is achieved. This would support a design dose of 5 mJ/cm^2 to achieve 5 log inactivation. Prior to 2003, the accepted UV dose for drinking water treatment application was 40 mJ/cm^2. Therefore at the June 11, 2002 progress review meeting, it was decided that MWRA would not attempt to prove UV doses lower than 40 mJ/cm^2 for disinfection purposes. In 2003, the proposed Long Term 2 Enhanced Surface Water Treatment Rule UV dose requirements were released. According to the LT2ESWTR, 2-log *Cryptosporidium* inactivation requires a dose of 5.8 mJ/cm^2 and 3-log inactivation requires a dose of 12 mJ/cm^2; these dose requirements account for uncertainty associated with the dose-response of the microorganisms. Hydraulic uncertainties and monitoring approaches require the addition of safety factors to the dose requirements, which could result in a UV dose of 21 mJ/cm^2 to 24 mJ/cm^2 for 2-log *Cryptosporidium* inactivation and 36 mJ/cm^2 to 42 mJ/cm^2 for 3-log *Cryptosporidium* inactivation according to the draft UVDGM released in June 2003.

The following sampling and analysis program was outlined in the original scope of the Ozone/UV project. There was continuous measurement of the flow to each train, raw water turbidity, raw water pH, raw water temperature, ozone system power use, and data from the on-line ozone analyzers. The instruments transmitted this information to the telog datalogger for storage. Periodically, the telog data was downloaded for data review and graphing.

Other data were recorded manually and entered into spreadsheets. These data included influent and effluent flows from the pipe loop racks, turbidity, temperature, chlorine residual, and pH values at the influent and effluents of the pipe loops for both trains, pressure readings at all gauges, calculated ozone dose from the gas feed monitor and the off gas monitor readings, and UV dose estimates from the UV unit control panels. Once a week, a graduate student from the University of New Hampshire was on-site to collect samples for total organic carbon and UV scan analyses at the University. Special disinfection by-product and bacterial samples were collected monthly and shipped to the University of Massachusetts and Montana State University. Once a month, MWRA personnel collected samples for determination of trihalomethane, haloacetic acid, iron and ammonia concentrations at the pipe loop influents and effluents. Twice during the course of the ozone/UV experiment, challenge trials of the primary disinfection systems using MS2 bacteriophage were performed. The protocol for these trials is outlined in the next section.

A compilation of the operational water quality data from the pipe loops after primary disinfection and post treatment obtained during the pilot study is presented in Table 4.2.

Table 4.2
Compilation of pilot plant water quality data – average values with standard deviations

Parameter	Raw water	UVTrain		Ozone/UV Train	
		Influent	Effluent	Influent	Effluent
Ozone dose (mg/L)	N/A	N/A	---	1.86 ± 0.90	---
pH	7.04 + 0.72	9.03 ± 0.35	9.19 ± 0.26	8.89 ± 0.39	9.06 ± 0.32
Alkalinity (mg/L as $CaCO_3$)	N/A	43 ± 10	45 ± 10	45 ± 9	47 ± 8.4
Total chlorine residual (mg/L)	N/A	2.18 ± 0.73	1.15 ± 0.55	2.22 ± 0.80	1.13 ± 0.78
Turbidity (NTU)	0.40 ± 0.04	0.44 ± 0.12	0.55 ± 0.52	0.48 ± 0.22	0.47 ± 0.20
Iron (mg/L)	---	0.038 ± 0.022	0.164 ± 0.046	0.040 ± 0.035	0.156 ± 0.053
Ammonia (mg/L)	---	0.17 ± 0.14	0.17 ± 0.14	0.20 ± 0.14	0.22 ± 0.16

There were no significant differences between Trains 1 and 2 with respect to influent alkalinity, total chlorine residual, temperature, iron or ammonia. There were also no significant differences between Trains 1 and 2 with respect to effluent alkalinity, total chlorine residual, temperature, iron, or ammonia. However, there were differences in the influent pH and turbidity values and the effluent pH and turbidity values between the two trains. Water temperature throughout the experiment varied seasonally as shown in Figure 4.2. The minimum water temperature occurred in mid January and the maximum water temperature peaked in mid August.

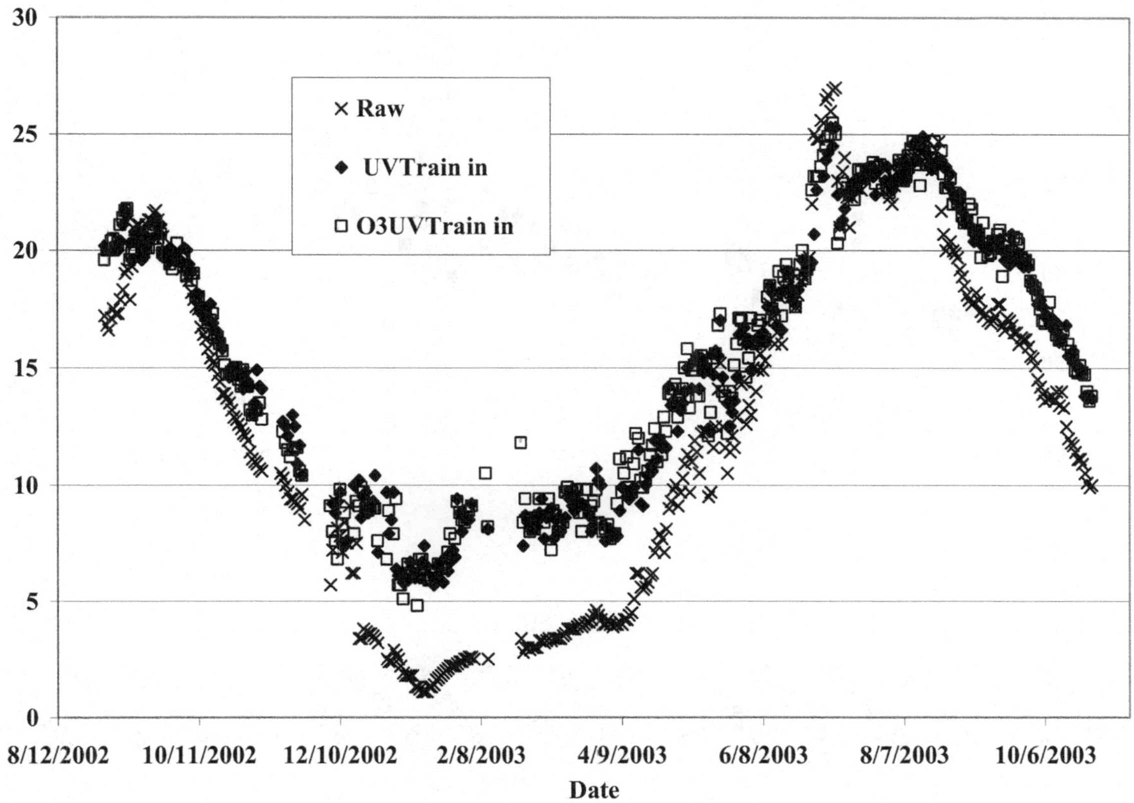

Figure 4.2 Water temperature throughout pilot trial

Weekly TOC and UV Scan Data

Samples were obtained weekly from the eight major sample locations of the pilot plant. Total organic carbon was analyzed in accordance with SM 5310 C. Persulfate – Ultraviolet Method. There were no significant differences in total organic carbon observed for the raw, ozone treated, UV-only treated or ozone/UV treated water samples as shown in Figure 4.3. Total organic carbon levels rose throughout the sampling period of just over a year. During the initial sampling in fall 2002, MWRA had transferred large quantities of higher quality Quabbin Reservoir water to supplement the Wachusett water during the drought of 2002. The Quabbin transfer decreased through the fall, stopping November 29, 2002. Through the winter and spring of 2003, the Wachusett Reservoir was recharged from watershed runoff and precipitation so the water quality reflected the predominance of lower quality, higher TOC Wachusett water with rising total organic carbon concentrations through most of the sampling period. Quabbin transfer started in early July 2003, later than usual due to ample precipitation, so the Wachusett water quality was less affected by the introduction of the higher quality Quabbin water than was the case in 2002.

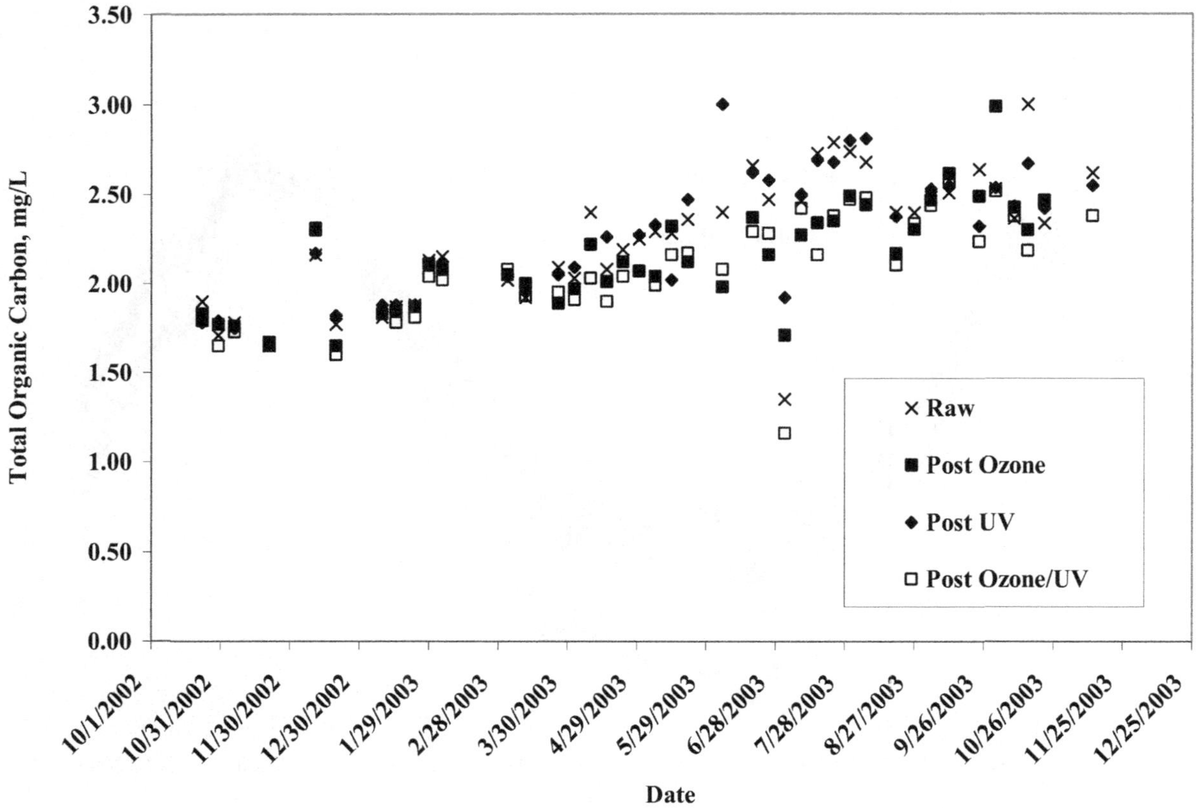

Figure 4.3 Total organic carbon data for raw, ozonated and UV irradiated water

Total organic carbon concentrations at the influents and effluents of the pipe loops were very similar to the concentrations in the raw water until May, when the effluent TOC concentration from the Ozone/UV Train Pipe Loops decreased and remained lower than the concentration of TOC at any of the other sample locations. The same trend of slightly rising TOC values throughout the trial period can be observed for the influent and effluents for the Pipe Loops in Figure 4.4.

The decrease in TOC observed as the ozonated, UV irradiated, and post treated water passed through the pipe loop from May onward may have resulted from the biodegradation of ozone oxidized organics by heterotrophic bacteria in the piping. Water temperature rose from 10 °C to 15 °C through the month of May, facilitating microbial growth in both of the Trains, but the non-oxidized organic matter present in the UV only Train may have been more resistant to microbial degradation, thus demonstrating no measurable decrease in TOC as the water was transported through the pipe loops while the Ozonated organic matter was assimilated.

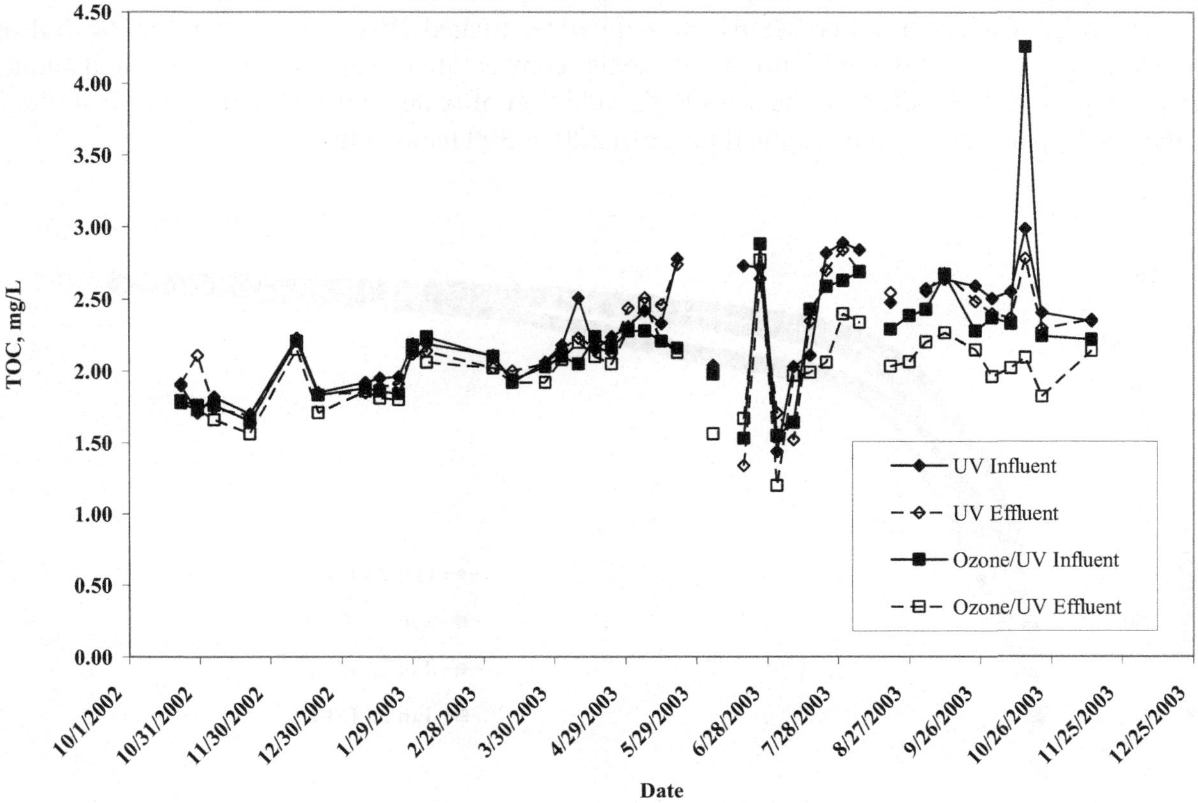

Figure 4.4 Total organic carbon data for UV Train and Ozone/UV Train pipe loops

Over the course of the Ozone/UV project, weekly UV scans of raw water, Train 1 water (non-ozonated, UV irradiation only) and Train 2 water (ozonated and UV irradiated) were conducted. The UV scan conducted on the samples is a modification of the SM 5910B Ultraviolet Absorbance Method where the scan varies over the range of ultraviolet wavelengths from 190 nanometers to 400 nanometers, rather than obtaining the absorbance at 254 nanometers as specified in the analytical method. The scan is preferable to the single wavelength data because medium pressure UV lamps are polychromatic, emitting at many wavelengths throughout the ultraviolet spectrum, and the scan can determine variations in UV absorbance in the water throughout this spectrum.

The use of ozonation prior to ultraviolet light disinfection of water can have two direct effects. Ozone is a powerful oxidant that acts as an electrophile, attacking the double bonds and activated ring structures in organic molecules found in the water. It is these double bonds and ring structures that also absorb UV light. Thus, ozonation can significantly increase ultraviolet light transmittance. This results in a more efficient transfer of UV energy into the water, making disinfection more efficient. On the negative side, ozone molecules also act as strong absorbers of UV photons, and if any ozone residual is present in the UV reactors, the transmittance of the water will be reduced, increasing the power requirements for disinfection.

Improvements in UV transmittance may be expected to correlate with ozone dose, ozone demand of the water, and water temperature. Figure 4.5 presents some comparative UV scans for the UV only Wachusett water versus the ozone/UV treated Wachusett water for the fall and winter sampling period while Figure 4.6 presents representative scans for the spring and summer sampling period. Regardless of the season, the addition of ozone improved the UV transmittance of the water, particularly in the critical range of 220 to 300 nanometers.

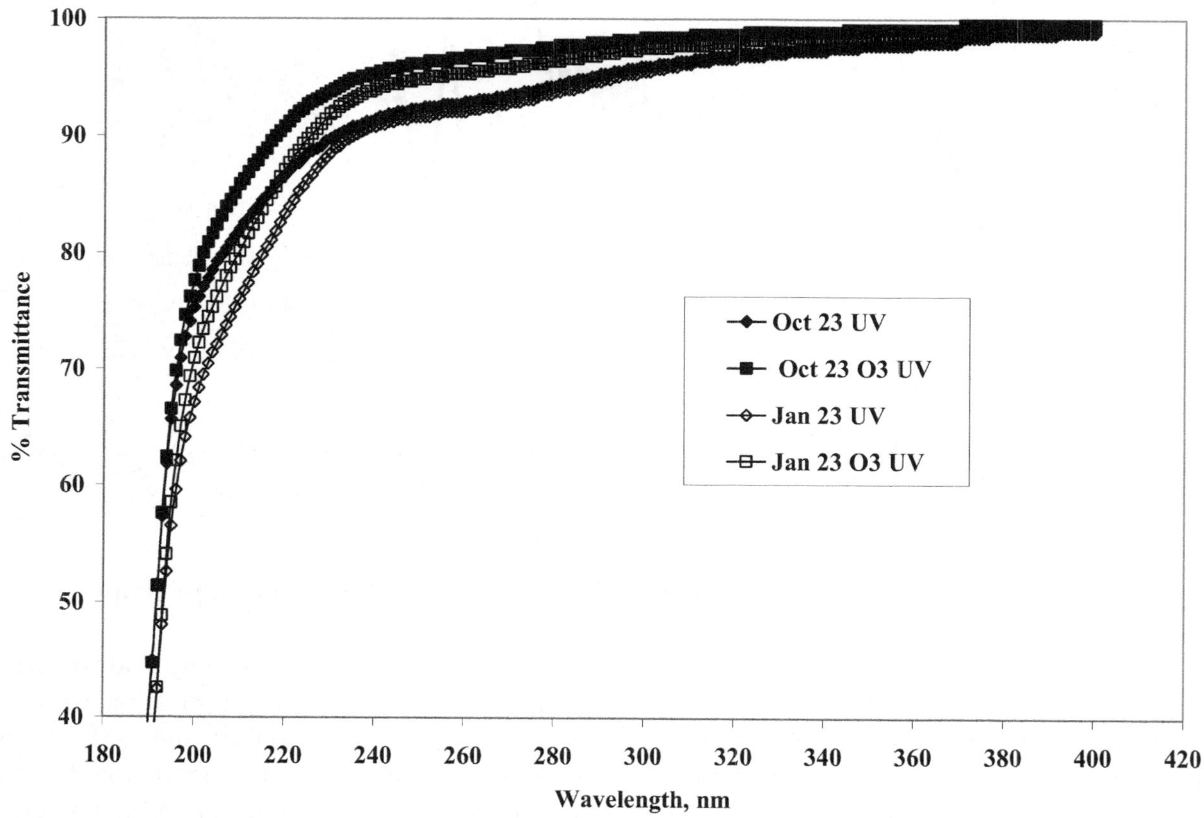

Figure 4.5 UV transmittance scans during the fall and winter of UV only and ozone/UV treated Wachusett water

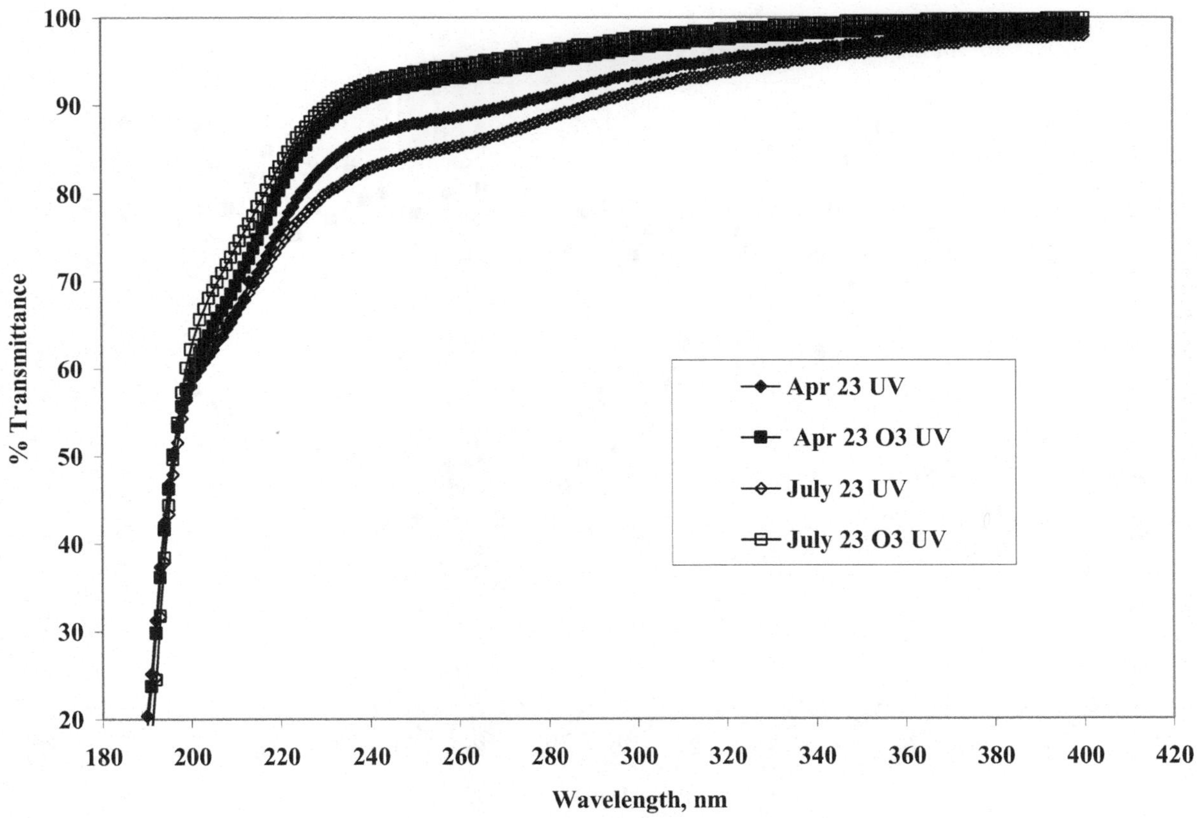

Figure 4.6 UV transmittance scans during spring and summer of UV only and ozone/UV treated Wachusett water

The ozone treated water demonstrated reduced UV absorbance compared to the raw water. Figure 4.7 presents the weekly summary of % UV transmittance at 254 nm of the raw, non-ozonated and pre-ozonated water. Using absorbance at 254 nm as an estimator for the difference in UV absorbance found for the ozonated versus non-ozonated water, Table 4.3 presents a summary of the UV transmittance of the raw, Train 1, and Train 2 water along with the ozone dose so that the effect of ozone on the transmittance of Wachusett water can be observed. The raw water sample was obtained before any treatment (sample point #1), the ozonated water was collected after ozonation and UV irradiation (sample point #3A) and the non-ozonated water was collected after UV irradiation (Sample point #3B). The ozone residual is dissipated by the time that the water passes through the UV unit of Train 2, but the oxidation of NOM by ozone increases the UV transmittance of the water by 2 to 9 percentage points, depending on initial water quality and ozone dose.

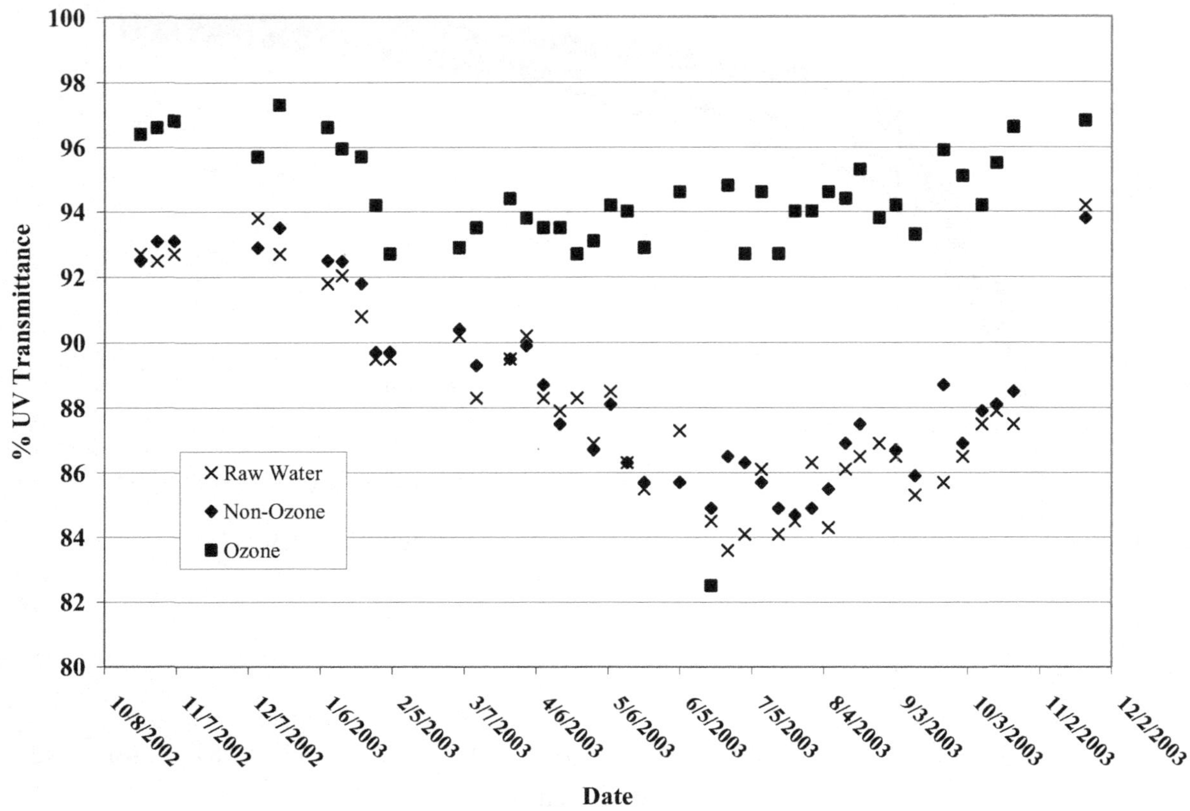

Figure 4.7 UV transmittance of raw, non-ozonated and pre-ozonated Wachusett water

Table 4.3
Summary of UV transmittance data for Wachusett water

	UV transmittance			Ozone dose (mg/L)
	Raw water (%)	Non-ozonated (%)	Ozonated (%)	
Average	88.1	88.5	94.5	1.96
Minimum	83.6	84.7	92.5	0.78
Maximum	94.2	93.8	97.3	3.23

There was a correlation of 0.888 between the ozone dose and the improvement in UVT transmittance,i.e., the higher the ozone dose, the greater the improvement in UVT. The upper limit of ozone dose at the MWRA pilot plant was approximately 3 mg/L and the relationship between ozone dose and improvement in UV transmittance was consistent in dosage range of 0 to 3 mg/L.

Challenge Trials

Field challenges combined with the collimated beam studies conducted during the bench scale inactivation work allow the calculation of a reliable UV dose for the pilot equipment. This dose is often referred to as the reduction equivalent dose (RED) in the literature because it is derived from a known reduction (inactivation) of the biodosimetry test organisms – MS2 bacteriophage in this case. The RED is used to track performance of the UV system through varying influent water qualities.

Materials and Methods

The general procedure used in field challenges with MS2 follows:

1. ~6 Liters of ~$10^{\wedge 10}$ plaque forming units/mL MS2 Stock is brought to the site.
2. Label all MS2 sampling vials (sterile 15 mL polypropylene centrifuge tubes).
3. Record Sample Number and Sample Description on MS2 Challenge Data Sheet.
4. Be sure there is no free and minimal total chlorine in the influent water by DPD or similar testing procedures.
5. Ensure UV systems to be tested are stable and running at the prescribed operating conditions.
6. Calibrate the MS2 feed pump to deliver the appropriate amount of MS2 stock solution (plus/minus 10%) necessary for the desired influent titer (100 mL/min is a typical flow rate).
7. Record Stock MS2 Concentration Reported on the Batch Shipped from the University of New Hampshire
8. Take a 15 mL sample of the MS2 stock for shipment to UNH as a QA/QC check.
9. If necessary, dilute the MS2 Stock with influent water to make the required volume for the duration of the test.
10. Allow this Spiking Mixture to mix for at least 30 minutes before use.
11. Take a 15 mL sample of the Spiking Mixture for shipment to UNH as a QA/QC check.
12. Take a 15 mL sample of the unspiked influent water for shipment to UNH as a QA/QC check.
13. Collect a 1 Liter Sample of unspiked influent water in a brown glass bottle for shipment to UNH for collimated beam testing.
14. Adjust pilot system sampling valves so a continuous flow of about 1 gpm will flow from the sampling taps (direct these to a floor drain or to a 5-gallon bucket.
15. Adjust pilot UV system flows to provide the desired rates.
16. Be sure all UV pilot system data are recorded and systems are running stable. Desired data is plant flow, dose and residual of any applied disinfectants, pH and temperature of the water, UV lamp power and sensor readings.
17. Start the Spiking Mixture feed and record needed information on the MS2 Challenge Data Sheet (using a laptop and Excel spreadsheet is recommended).
18. Allow Spiking Mixture to run for 15 minutes before sampling.
19. Take all samples needed- recommended that samples be taken in 3 rounds of influent and effluent spaced 15 minutes apart.
20. Record END of RUN information for both UV pilot systems.
21. Shut off UV lamps on all pilot systems, CONTINUE TO RUN MS2 spike.
22. After 15 minutes, sample effluent locations from each UV system as the final QA/QC Samples Taken.
23. Spike pump can now be shut off and spiking system flushed with influent water to clean it out.

24. Thoroughly clean all spiking containers, pump and tubing at sampling locations in preparation for next run.
25. Pack all 15 mL samples and the 1 L sample carefully to avoid breakage, add blue ice packs and ship for next day delivery to the UNH.

First Challenge Trial

The first MS2 challenge trial was conducted on October 9, 2002 at the Pipe Loop Pilot Facility. The raw water quality was as follows; pH 6.71, temperature 17.7 °C, turbidity 0.39 NTU. The first part of the trial was conducted using Train 1, the UV only train, to determine the RED for the Wachusett water. Three UV lamp power levels were tested; 100%, 67.3%, and 30%. In the second part of the trial, the MS2 stock was applied to Train 2 after ozone application/contact but before the Train 2 UV unit.

The results of MS2 challenge on Train 1 are shown in Table 4.4. The results shown in Table 4.4 are predictable and indicate that the UV reactor was responding as expected to changes in flow and/or lamp power. A thorough examination of the data also indicates that replicate samples of the influent and effluent for any given condition show very good precision. In addition, all QA/QC sample results (e.g., trip blanks, sterile controls, background field blank MS2 samples and samples taken with the UV lamps off) produced the expected results, indicating that the testing was controlled and meaningful data resulted.

Table 4.4
Biodosimetry results for the UV reactor in Train 1 (no ozone) – October 9, 2002

Flow (gpm)	Lamp power setting (%)	Mean influent MS2 (log)	Mean effluent MS2 (log)	Mean MS2 inactivation (log)	Calculated RED* (mJ/cm^2)
400	100	6.63	3.38	3.25	72.2
400	67.3	6.63	4.23	2.40	51.9
400	30	6.63	5.57	1.06	19.9
200	67.3	6.92	3.63	3.29	73.2
200	30	6.92	5.35	1.57	32.0
200	0	6.98	6.90	0.08†	NA

* Calculated from the collimated beam data using a linear regression analysis with the resulting equation: log MS2 Inactivation = 0.0419 (RED) + 0.224; r-squared = 0.994
† Indicates good QA/QC and no significant unexplained losses of MS2

Experimental results from Train 1 clearly show that the UV reactor is performing as expected and is responsive in a predictable manner to changes in flow and changes in lamp power.

Unfortunately, experiments on Train 2 did not produce significant results for two reasons. First, it was not possible to increase the Train 2 flow to higher than 212 gpm; therefore, the experimental plan was modified to use two different ozonation doses (~1.85 mg/L and ~1.5 mg/L) at the flow rate of 200 gpm. Second, for the higher ozone dose case, it is obvious in examining the raw MS2 counts (all BDL - zero plaques on the plate counts actually) from Train 2 experiments, particularly the influent samples, that ozone was getting through to the UV influent at levels sufficient to eliminate all of the MS2 added to the system. This was not expected since the pilot ozone system probes suggested that the ozone residual was below detection leaving the fifth contactor of six. Grab samples tested using the HACH-Indigo method also suggested that the ozone levels were BDL. However, it is quite possible that the detection limits for both the ozone probes and the high range HACH indigo ampules were high enough (insensitive enough) to fail to detect a residual of 0.05 mg/L. Unfortunately, MS2 is highly susceptible to inactivation by even these low concentrations of ozone.

In the lower ozone level trials on Train 2 there is survival of MS2 in the influent samples, with levels reaching about 10^4 to 10^5 pfu/mL (as opposed to the 10^7 pfu/mL levels expected and observed in Train 1 influent during testing). However, at a power level of 67.3% the effluent MS2 levels were reduced to below detection (zero plaques on the plate counts actually). The 30% UV power level data for the lowest ozone dose did produce influent and effluent MS2 counts. From these data, the RED was found to be 50.6 mJ/cm^2. If we compare this to line 5 of Table 4.3 above under similar testing conditions, it might suggest that the use of ozone prior to the UV reactor increased the calculated RED from 32 mJ/cm^2 to 50.6 mJ/cm^2 which would be significant. However, the data are suspect since there is evidence that ozone was getting to the UV reactor based on the reduced MS2 influent counts, so it is likely that the MS2 inactivation is actually the result of both ozone and UV thus making the calculated RED for the UV system unrealistically high. Experimental results from Train 2 suggest that ozone may be improving the UV performance, but the data are suspect and limited.

Second Challenge Trial

The second MS2 challenge trial was conducted on June 5, 2003 at the Pipe Loop Pilot Facility. The raw water quality was as follows: pH 6.41, temperature 14 °C, turbidity 0.35 NTU. The first part of the trial was conducted using Train 1, the UV only train, to determine the reductive equivalent dose for the Wachusett water. Three UV lamp power levels were tested; 100%, 67.3%, and 30%. In the second part of the trial, the MS2 stock was applied to Train 2. Unfortunately, as a power spike occurred the morning of the trial, resulting in the loss of the ozone generator, no ozone could be applied to the water of Train 2, so the trial was used to compare the performance of the two units rather than to determine any benefits of pre-ozonation for UV disinfection.

The results of MS2 challenge on Train 1 are shown in Table 4.5. The results shown in Table 4.5 are predictable and indicate that the UV reactor was performing as expected to changes in flow and/or changes in lamp power as well as showing the effect of aging on lamp performance. A thorough examination of the data also indicates that replicate samples of the influent and effluent for any given condition show very good precision. In addition, all QA/QC sample results (e.g., trip blanks, sterile controls, background field blank MS2 samples and samples taken with the UV lamps off) produced the expected results indicating that the testing was controlled and meaningful data resulted.

Table 4.5
Biodosimetry results for the UV reactor in Train 1 (no ozone) – June 5, 2003

Flow (gpm)	Lamp power setting (%)	Mean influent MS2 (log)	Mean effluent MS2 (log)	Mean MS2 inactivation (log)	Calculated RED* (mJ/cm^2)
435	100	6.85	4.93	1.92	46
435	48.7	6.85	5.75	1.10	26.4
435	30	6.85	6.11	0.74	17.4

* Calculated from the collimated beam data using a linear regression analysis with the resulting equation: log MS2 Inactivation = 0.0416 (UV dose) + 0.0001, r squared = 0.995

The effect of lamp aging can be observed for the higher lamp power settings. There is only a minor difference in calculated RED between June 5 and October 9 for the 30% lamp setting, but a decrease of 26 mJ/cm^2 was observed for the 100% lamp settings when the earlier calculations are compared with the later.

The results of MS2 challenge on Train 2 are shown in Table 4.6. The results shown in Table 4.6 are predictable and indicate that the UV reactor was responding as expected to changes in flow and/or lamp power. As the data from the October 9 challenge were compromised by ozone carryover, no evaluation of UV performance over time for the UV unit for Train 2 is possible. It is interesting to note that the calculated reduction equivalent dose values for Train 2 are similar to those of the Train 1 UV unit.

Table 4.6
Biodosimetry results for the UV reactor in Train 2 (no ozone) – June 5, 2003

Flow (gpm)	Lamp power setting (%)	Mean influent MS2 (log)	Mean effluent MS2 (log)	Mean MS2 inactivation (log)	Calculated RED* (mJ/cm^2)
308	100	7.04	5.12	1.92	46.3
270	48.7	7.04	6.01	1.03	24.7
270	30	7.04	6.47	0.57	14.0
150	30	7.04	6.45	0.59	14.4
308	0	7.04	6.97	0.07	N/A

* Calculated from the collimated beam data using a linear regression analysis with the resulting equation: log MS2 Inactivation = 0.0416 (UV dose) + 0.0001, r squared = 0.995

Monthly Disinfection By-Product Data

Once a month, paired samples were collected from each pipe loop influent to determine the disinfection by-product concentrations for each treatment train. Paired samples were also collected and analyzed from the pipe loop effluents to observe the effect of interaction within the pipelines on disinfection by-product formation and decay. For some of the disinfection by-products, samples were collected of the raw water, of the water from Train 2 after ozone application and from the paired treatment trains after UV irradiation to determine if there were differences throughout the treatment process. For the paired data, comparisons were made using the t-test at a 95% confidence level. Disinfection by-products studied were as follows:

- Trihalomethanes
- Haloacetic acids
- Total Organic Halides (TOX) with size fractionation of the organic halides
- Aldehydes
- Keto-Acids

Trihalomethances

The trihalomethanes (THMs) and haloactic acids, common by-products of chlorination, are regulated drinking water contaminants with maximum contaminant levels (MCLs) of 0.080 and 0.060 mg/L, respectively. Total organic halides present information about a range of halogenated organic materials, both regulated and currently unregulated, resulting from chlorination or oxidation processes, Aldehydes and keto-acids are oxygenated by-products commonly created by ozonation of the water, and both are easily broken down by heterotrophic bacteria (Kitchens et al, 1976).

Sample collection for evaluation of disinfection by-products was initiated during the December 2002 sampling and continued monthly through November 2003, except for the month of February when the raw water line froze. As the water temperature increased through the months of May through August, so did the levels of trihalomethanes in the pipe loop influents and effluents as shown in Figure 4.8. As the water temperature and TOC levels increased, the concentration of trihalomethanes in Train 1 diverged from the concentrations observed for Train 2, with lower levels of trihalomethanes observed for the pre-ozonated water both at the influent to the pipe loops (after chlorine contact) and the effluent of the pipe loops. A graph showing the major halogenated disinfection by-product concentrations (THM, HAA6, and TOX) versus water temperature is presented as Figure 4.12.

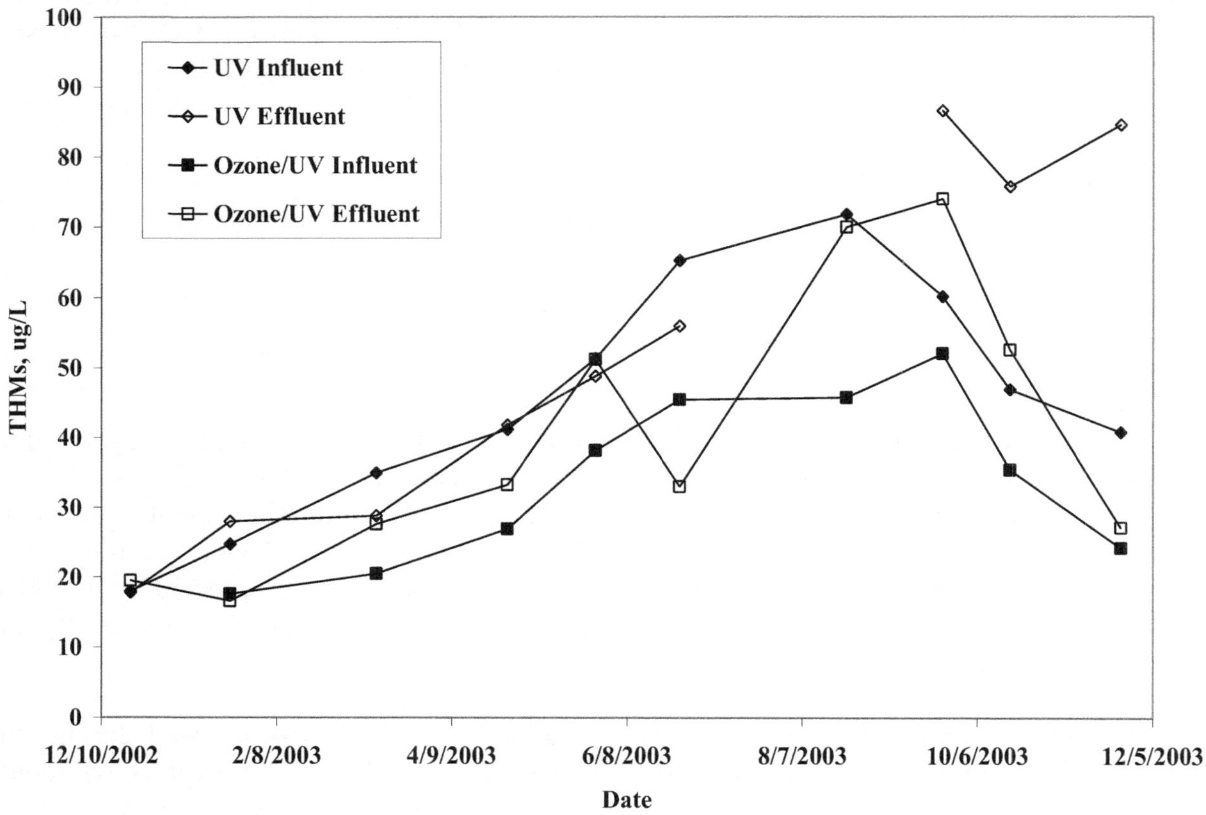

Figure 4.8 Trihalomethane data for UV Train and Ozone/UV Train pipe loops

Haloacetic acids

Haloacetic acid samples for HAA5, the regulated haloacetic acids, and HAA6 were obtained at the same time and in the same locations as the trihalomethane samples. The same trend showing increasing HAA5 levels with increasing water temperature during the late spring and summer months of May through September be clearly observed. There were differences in HAA5 concentrations determined between the trains for both the paired influent and paired effluent samples. These data are presented in Figure 4.9.

The HAA6 samples add the results of bromodichloroacetic acid analysis to the haloacetic acids determined for the HAA5 analysis. With the addition of the bromodichloroacetic acid data, the overall HAA6 concentrations were still determined to be different between the two trains, with the pre-ozonated Train 2 concentrations consistently less than the non-ozonated Train 1 concentrations at the influents and effluents of the pipe loops. These data are presented in Figure 4.10.

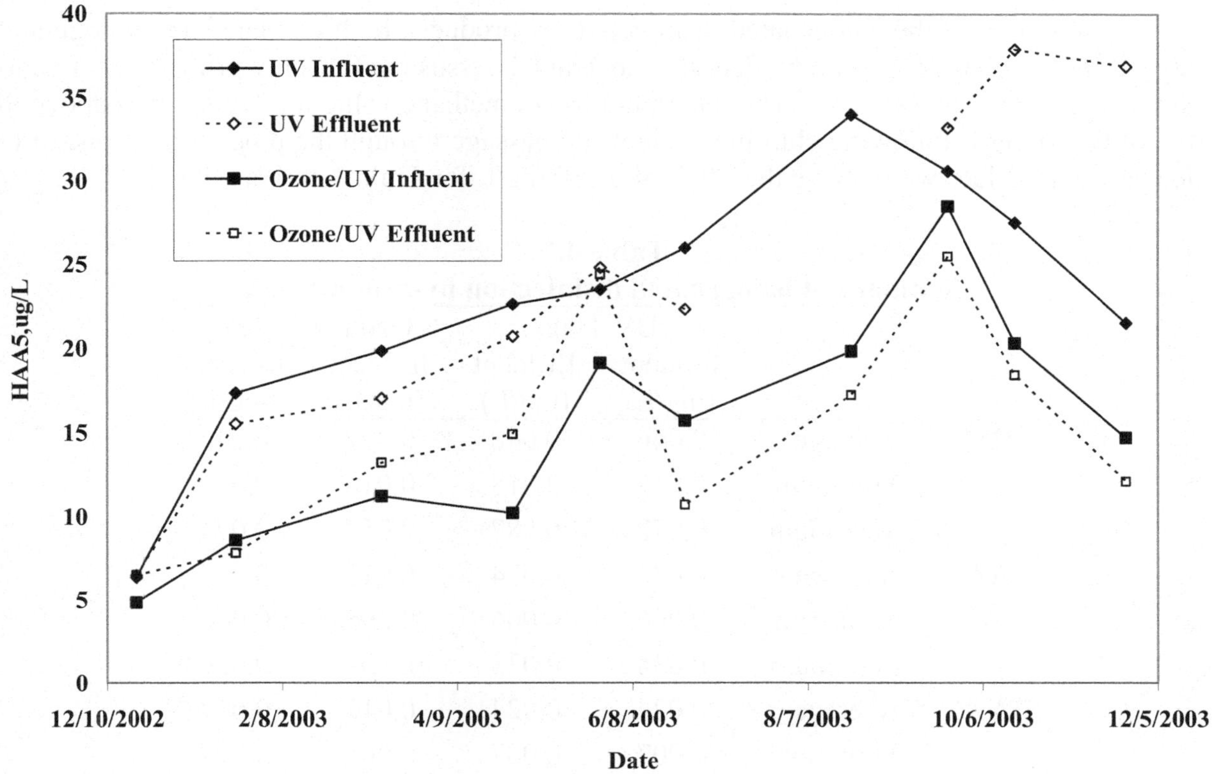

Figure 4.9 Haloacetic acid 5 data for UV Train and Ozone/UV Train pipe loops

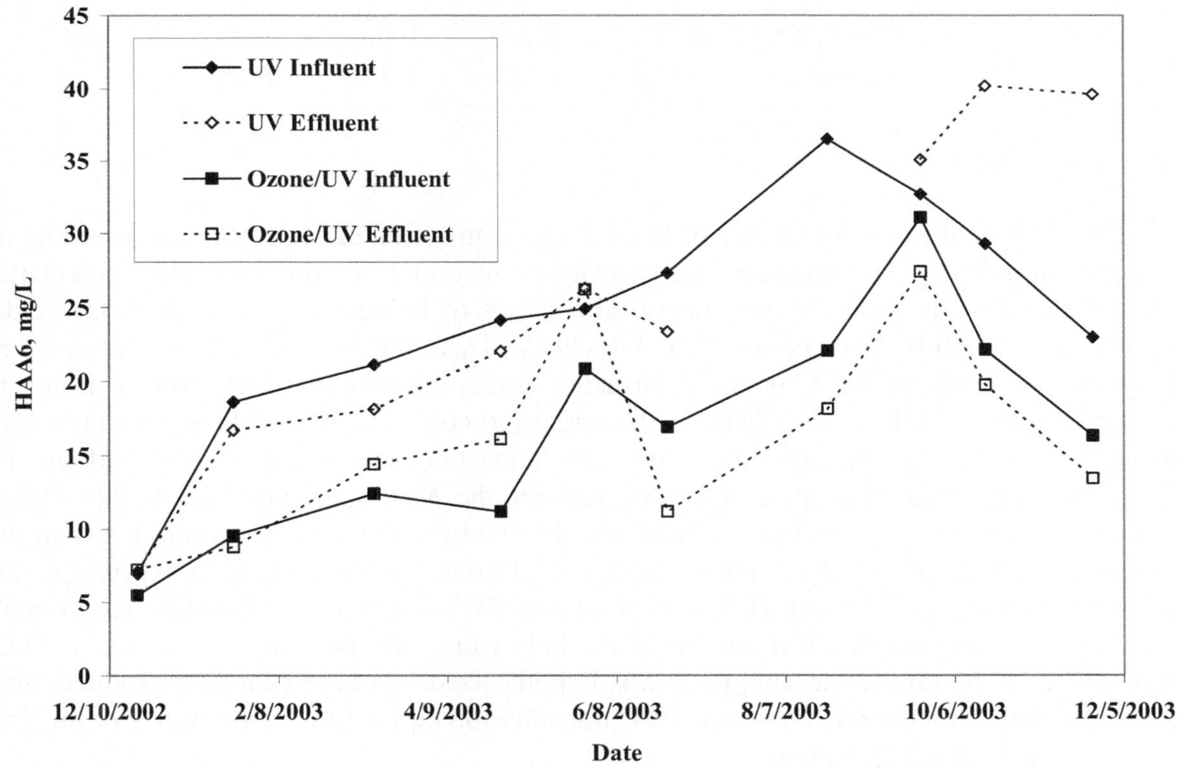

Figure 4.10 Haloacetic acid 6 data for UV Train and Ozone/UV Train pipe loops

A summary of the halogenated disinfection by-products, both regulated and unregulated, analyzed for the ozone/UV primary disinfection Train 2 versus the UV only primary disinfection Train 1 is presented in Table 4.7. The maximum trihalomethane value for Train 1 was above the MCL of 0.080 mg/L following chlorine contact and passage through the pipe loops. None of the haloacetic acid values were above the MCL of 0.060 mg/L for Train 1 or Train 2.

Table 4.7
Summary of halogenated disinfection by-product data

		UV Train		Ozone/UV Train	
		Influent (mg/L)	Effluent (mg/L)	Influent (mg/L)	Effluent (mg/L)
THMs	Average	0.046	0.048	0.035	0.042
	Minimum	0.018	0.018	0.018	0.017
	Maximum	0.072	0.087	0.052	0.074
HAA5	Average	0.023	0.024	0.015	0.015
	Minimum	0.006	0.006	0.005	0.006
	Maximum	0.034	0.038	0.0285	0.026
HAA6	Average	0.024	0.024	0.017	0.017
	Minimum	0.007	0.007	0.005	0.007
	Maximum	0.0365	0.040	0.031	0.027
TOX	Average	0.116	0.142	0.112	0.128
	Minimum	0.075	0.096	0.081	0.089
	Maximum	0.172	0.188	0.137	0.163

Total organic halides

The TOX analysis is a measure of the full spectrum of halogenated organic compounds that are present in water. This includes well known disinfection by-products like trihalomethanes and haloacetic acids, as well as other many other types of halogenated acids, hydroxy-acids, aldehydes and keto-acids (Venkatesan, N. and Reckhow, D.A,. 2003). TOX data are presented in Figure 4.11. The levels of TOX for the chlorinated waters introduced to and flowing from the pipe loops generally follow the pattern of increasing concentration with increasing water temperature observed for the trihalomethane and haloacetic acids samples analyzed by the MWRA laboratory, though a spike was observed for the May sampling data at the UMass laboratory. There were no significant differences observed for TOX concentration between the influents of the UV treated train when compared to the Ozone/UV treated train, however, the UV treated water had higher levels of TOX than the ozone/UV treated water after passage through the pipe loops. During the first few months of the field work, unexpectedly high levels of TOX were observed in the raw water sample. It was hypothesized that chlorinated by-products may have been leached from the relatively new PVC transmission piping to the bulk water for the first few months of pilot plant operation.

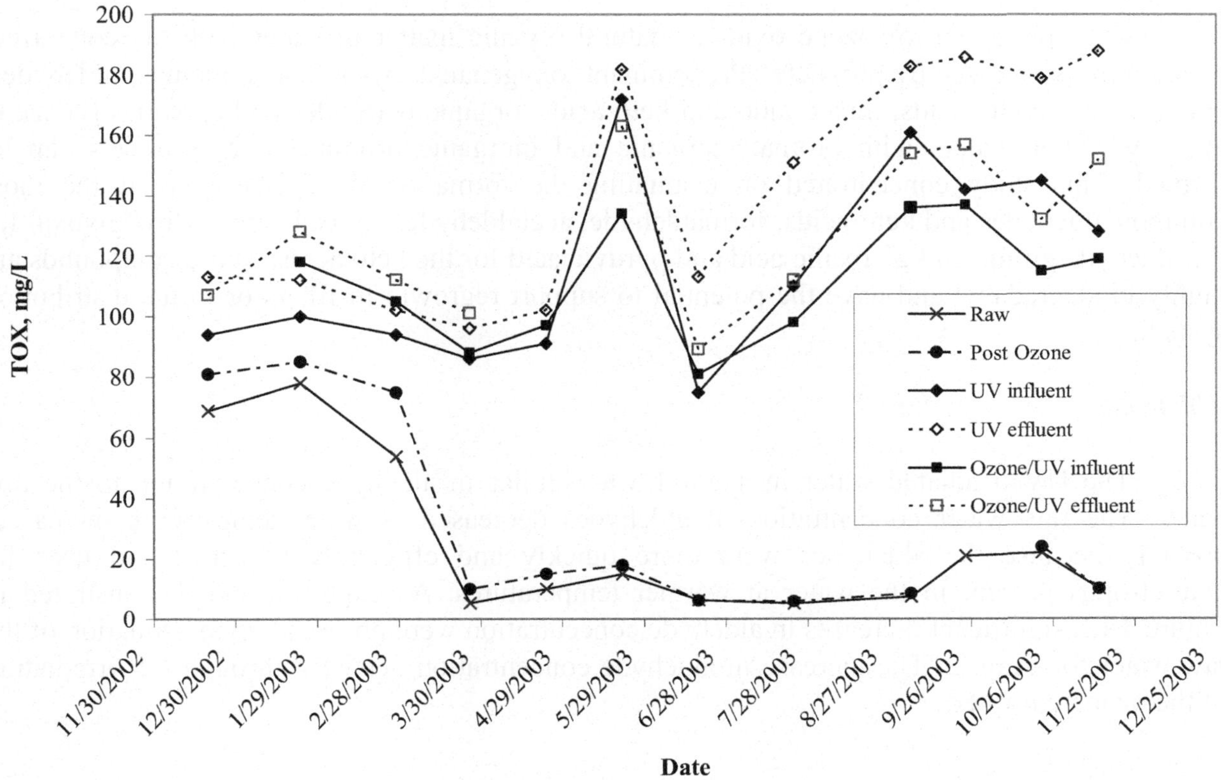

Figure 4.11 TOX data for the UV Train and Ozone/UV Train pipe loops

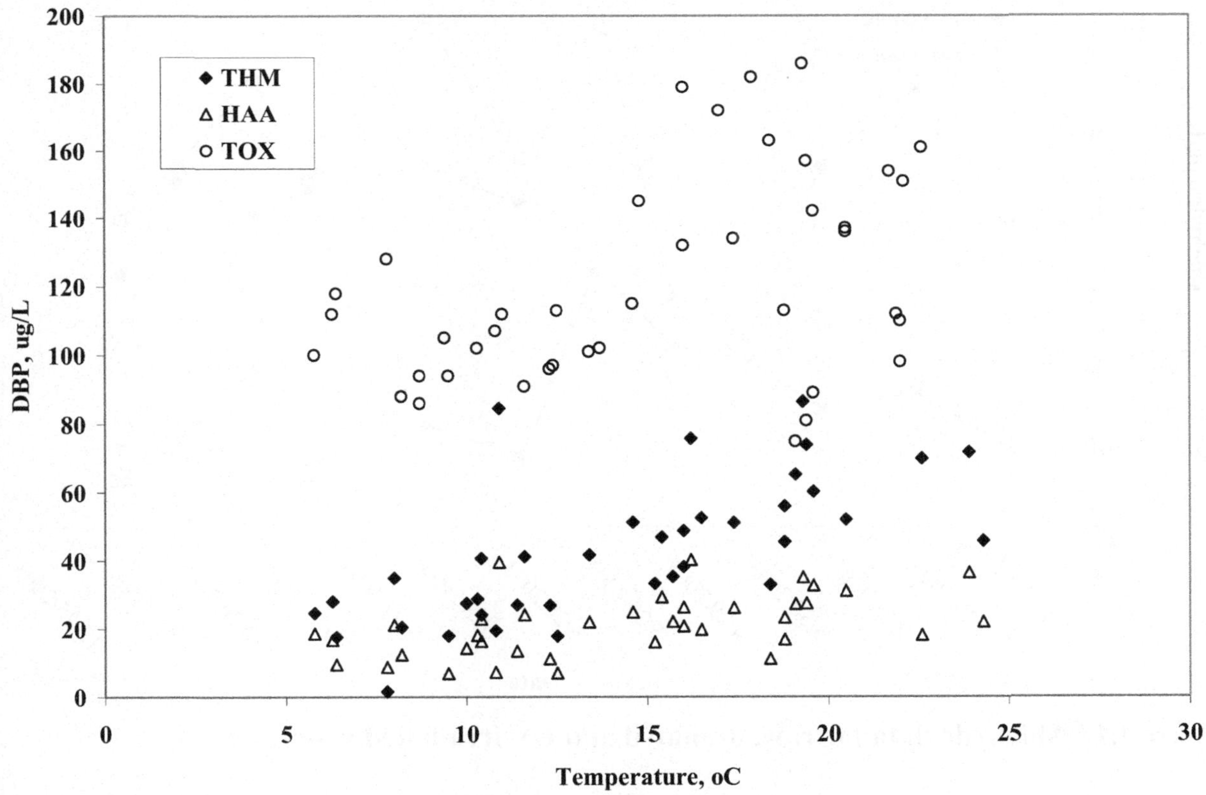

Figure 4.12 Halogenated DBP versus Water Temperature

The application of ozone oxidizes natural organic matter in water with the concurrent creation of oxygenated by-products. Predominant oxygenated by-products include aldehydes, ketones, carboxylic acids, mixed aldo and keto acids, organic peroxides and epoxides (Glaze et al., 1993). For waters with bromate, organic and inorganic brominated by-products can be formed. This study concentrated on evaluating the formation and persistence of the most common aldehydes and keto-acids; formaldehyde, acetaldehyde, glyoxal, and methyl glyoxal for the aldehyde group, and glyoxalic acid and pyruvic acid for the keto acids. These compounds are highly biodegradable and have the potential to support regrowth in filters or in the distribution system.

Aldehydes

The UV irradiated water in Train 1 was similar in aldehyde concentration to the raw water. The raw water concentration of aldehydes decreased as water temperature increased, possibly because the aldehydes were more quickly and efficiently adsorbed and used by heterotrophs present in the water at warmer temperatures. As expected and demonstrated in Figure 4.13, significant increases in aldehyde concentration were observed upon ozonation of the raw water for Train 2. The increase in aldehyde concentration persisted through UV irradiation of the ozonated water.

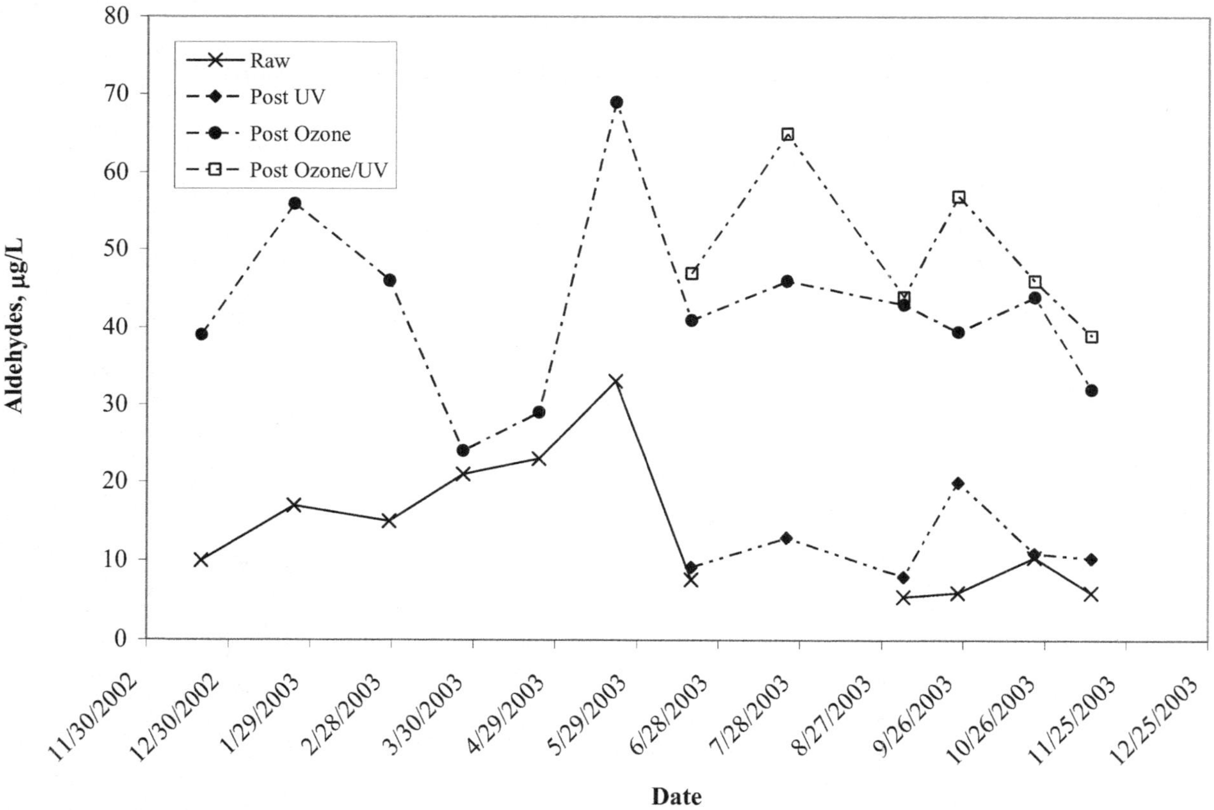

Figure 4.13 Aldehyde data for raw, ozonated and UV irradiated water

As the waters from both trains were chlorinated, stabilized through adjustment of pH and alkalinity, and chloraminated, the pre-ozonated, UV irradiated water retained higher levels of aldehydes compared to the UV irradiated water, as shown in Figure 4.14.

However, as the waters passed through their respective pipe loops, the levels of aldehydes in the UV irradiated water increased to levels that were not significantly different from the level of aldehydes in the pre-ozonated, UV irradiated water. Both chlorination and chloramination have been demonstrated to create aldehydes, so it is likely that the aldehydes in the UV pipe loops influents and effluents were formed during secondary disinfection. This would be supported by the fact that the UV irradiated pipe loop influent had higher concentrations of aldehyde than the raw or post UV water. Overall aldehyde levels decreased with increasing water temperature.

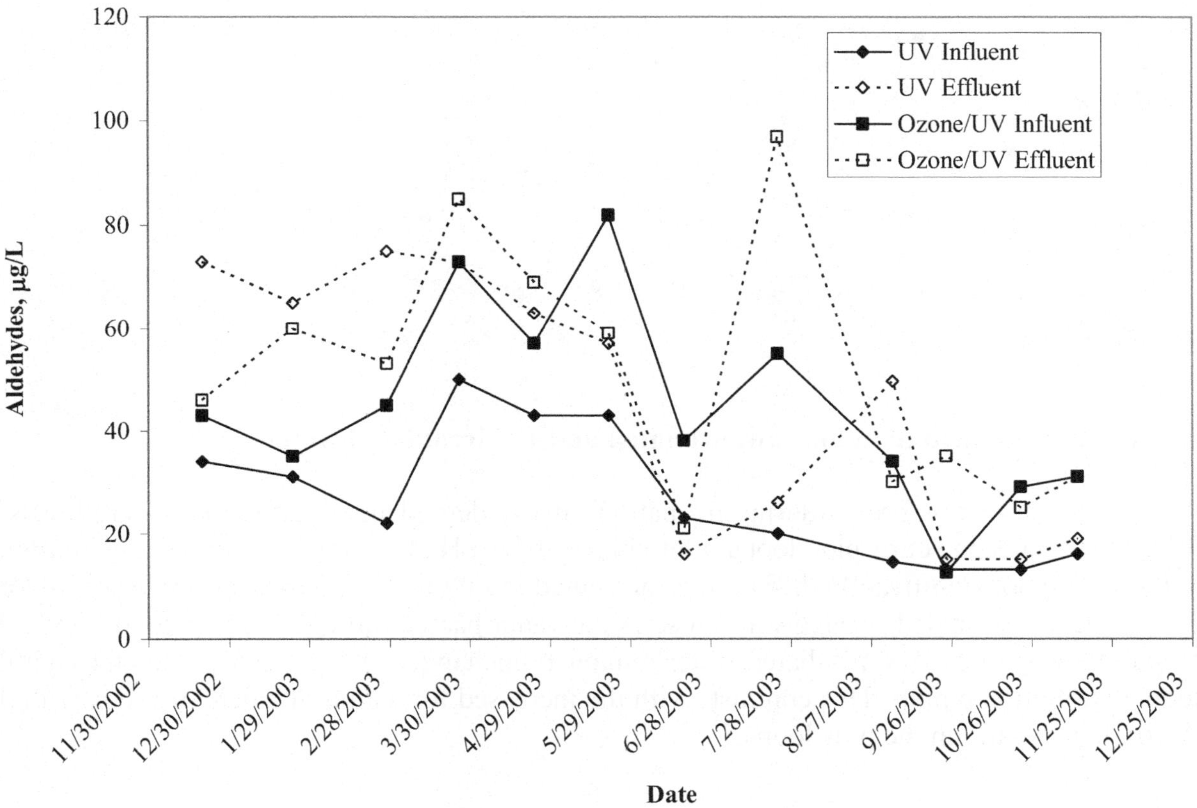

Figure 4.14 Aldehyde data for UV Train and Ozone/UV Train pipe loops

Keto-acids

As with aldehydes, the UV irradiated water in Train 1 was similar to the raw water with respect to keto-acid concentrations. Also similar to the observations with aldehydes, significant increases in keto-acid concentrations were observed upon ozonation of the raw water for Train 2, particularly through the warmer water months of May, June, July, and September. The increase in keto-acid concentration persisted through UV irradiation of the ozonated water. Figure 4.15 presents the raw, ozonated and post UV keto-acid results for the two treatment trains.

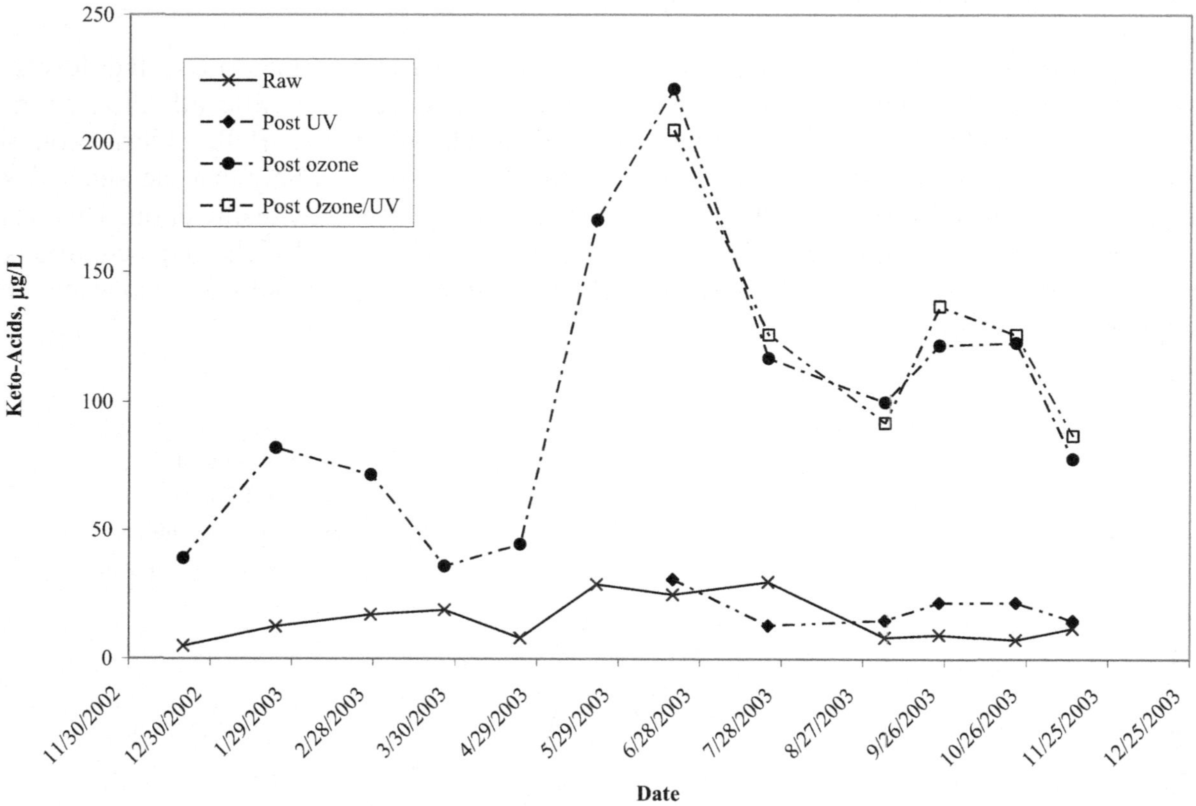

Figure 4.15 Keto-acid data for raw, ozonated and UV irradiated water

The keto-acids formed during ozonation rapidly degraded so that the keto-acid levels at the influents to each set of pipe loops after chlorination, pH/alkalinity adjustment and ammonia addition were not significantly different, as presented in Figure 4.16. However, the levels of keto acids in the pre-ozonated, treated water rose as the water passed through the pipe loops while the keto-acid levels in the UV irradiated water remained unchanged as the water passed through the model distribution piping. This contrasts with the increased formation of aldehydes seen with the UV irradiated water in the pipe loops.

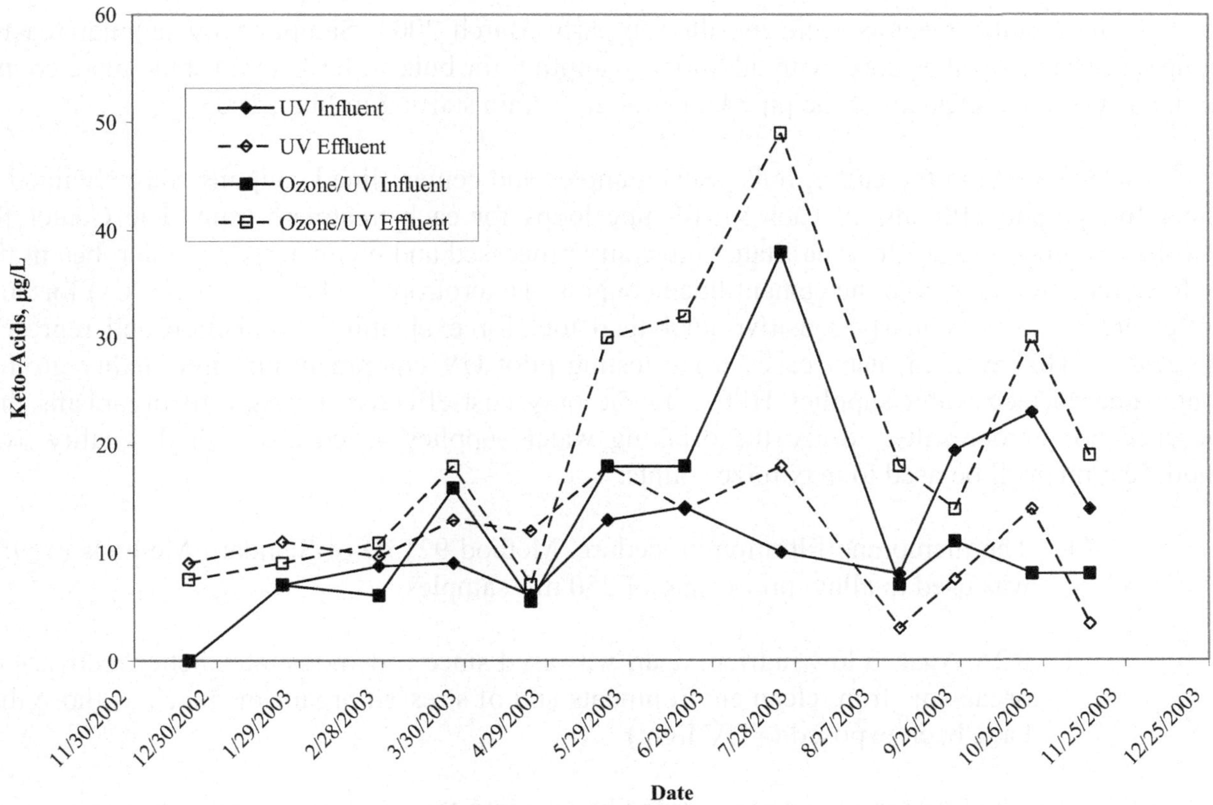

Figure 4.16 Keto-acid data for UV Train and Ozone/UV Train pipe loops

A summary of oxygenated disinfection by-products is presented in Table 4.8.

Table 4.8
Summary of oxygenated disinfection by-product data

		Raw (µg/L)	Post ozone (µg/L)	UV Train Influent (µg/L)	UV Train Effluent (µg/L)	Ozone/UV Train Influent (µg/L)	Ozone/UV Train Effluent (µg/L)
Aldehydes	Average	14	42	27	46	45	51
	Minimum	5.5	24	13	15	12	21
	Maximum	33	69	50	75	82	85
Keto-acids	Average	15	129	11	11	12	20
	Minimum	5	87	0	3	0	7
	Maximum	30	205	23	18	38	49

Monthly Regrowth Data

The annular reactors were installed in early March 2003. Sampling for annular reactor coupons started April 3, 2003 with additional sampling for bulk water heterotrophic plate counts at the influent and effluent of the pipe loops for each train starting in May, 2003.

Once a month thereafter, bulk water samples and cement lined coupons were obtained at the influents and effluents of each set of pipe loops for each treatment train. The Center for Biofilm Engineering at Montana State University processed and enumerated the microbes in the bulk water and attached to the cement lined coupons. Heterotrophic plate counts (HPCs) for bulk water are not a particularly sensitive analytical tool for evaluating disinfection and regrowth processes. However, in many cases when testing pilot UV equipment on high quality ground water and surface water supplies HPCs are the only cost-effective measure of organisms that produce non-zero results. Since the drinking water supplies tested are of high quality, two modifications will be used to maximize counts:

1. The membrane filtration procedure, Method 9215D of Standard Methods (1998), was used to allow processing of 250 mL samples.

2. R2A Agar, a low nutrient agar, was used since it demonstrates better recovery of organisms from clean environments and of stressed organisms (such as those that have been exposed to UV light).

The attached bacteria were enumerated by scraping the 13 square centimeter coupon face with a sterile loop to remove the biofilm, homogenizing the biofilm scrapings in sterile dilution water, diluting the homogenate and plating the solution using the Standard Method 9215C Spread Plate Method.

The attached heterotrophic bacterial counts from the annular reactor coupons are presented in Figure 4.17. There were no significant differences between the Trains 1 and 2 with respect to influent attached bacterial counts, nor were there differences between the effluent coupon counts from the UV Train and the Ozone/UV Train. For both trains, the effluent attached bacterial counts were significantly higher than the influent counts by approximately two orders of magnitude. This is not unexpected, given the loss of chloramine residual during passage of the water through the pipe loops.

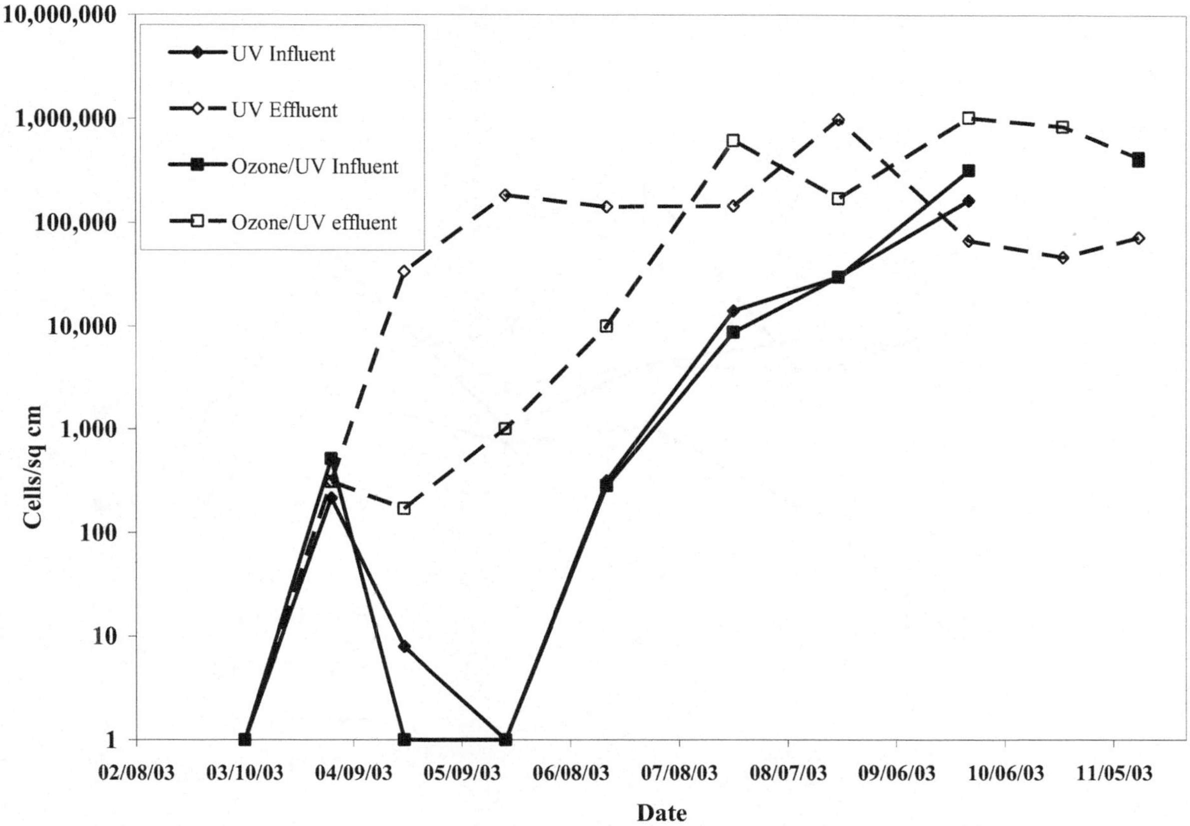

Figure 4.17 Attached HPC data for UV Train and Ozone/UV Train pipe loops

The bulk water had significant numbers of heterotrophs, in spite of an average influent total chlorine residual of 2 mg/L as shown in Figure 4.18. There is a significant difference between the UV illuminated influent water bacterial numbers and the ozone treated, UV illuminated influent water bacterial numbers, with the UV only HPC counts on average an order of magnitude higher than the ozone/UV HPC counts. Microbial numbers peaked in July and August with the warmest water temperatures though the numbers began to decrease only slowly as the water temperature lowered through September and October. Attached HPC do not rise significantly as the water passes through the pipe loops for the UV only train, but heterotroph numbers do increase with passage through the pipe loops for the ozone/UV train. Nevertheless, given the range of the data, the pipe loop effluent heterotroph numbers from the two trains were not statistically different for the duration of the testing.

There were not enough data points to perform correlations between HPC numbers and levels of the oxygenated disinfection by-products. The keto-acid concentration in the ozonated water decreased by 80 to 90% in approximately 3.5 hours, the time that it took the water to travel from the ozone contact tank outlet through the UV unit and chlorine contact piping to the inlet of the pipe loops. Given the presence of high levels of this very biodegradable carbon source, one may have expected that the pipe loops that received ozonated water would support higher levels of heterotrophic bacteria but that could not be proven in this pilot trial.

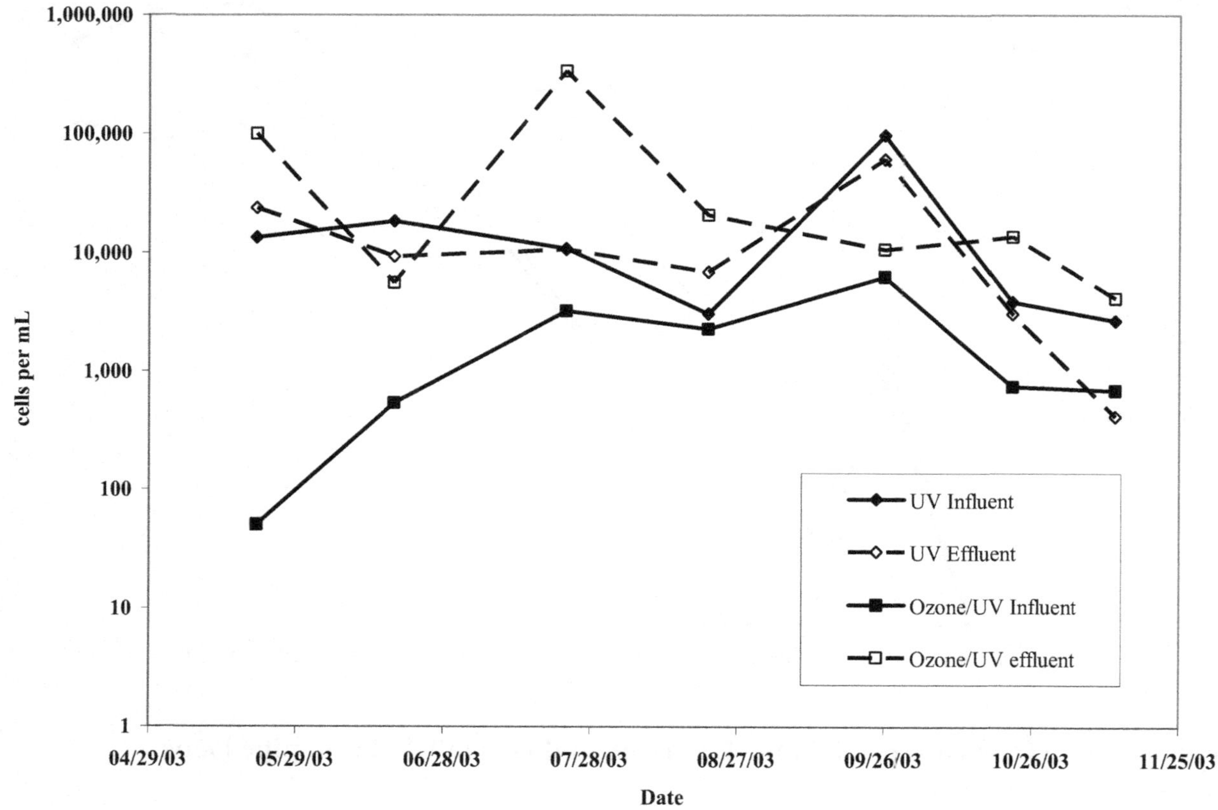

Figure 4.18 Bulk water HPC data for UV Train and Ozone/UV Train pipe loops

It is likely that the creation of high levels of oxygenated disinfection by-products via ozonation did result in lower levels of halogenated disinfection by-products in the ozone/UV Train, possibly by oxidation of potential halogenated precursors to aldehydes and ketones before the chlorine could react to form THMs or HAAs. Aldehydes and keto-acids can react with chlorine to form halogenated compounds but the reactions involved are complex and slow compared to halogenation of methyl groups. The reaction time available in the chlorine contact column (3.2 hours) and the pipe loops (3.7 hours) may not have been sufficient to observe much halogenation of oxygenated compounds particularly as this process was complicated by the competing loss of aldehydes, ketones and keto-acids as microbial carbon sources.

CONCORD FIELD TRIAL

The Town of Concord is one of the older and most historic suburbs of Boston. Its current population is approximately 15,000. The municipal water system servicing this community was established in 1872 and serves over 5,300 individual accounts. The customer base is predominantly residential; however, there is a significant mix of commercial and institutional users as well. Until 1949, the source of supply was exclusively from a surface water reservoir. Since then, the system has evolved to a combination of six groundwater wells and one surface water supply. The Nagog Pond source is chlorinated and pumped into the distribution system at the Route 2A Pump Station facility. In 1991, in response to stricter microbial inactivation requirements imposed by the Surface Water Treatment Rule, the Town of Concord constructed the Nagog Pond Ozone Facility. The Nagog Pond Ozone Facility, designed to provide 2 log inactivation of *Giardia lamblia*, went on-line in 1996. Raw Nagog Pond water quality is such that the Town of Concord has been able to maintain a Waiver to Filtration since 1991.

A full scale trial of ozone treatment followed by medium pressure UV was conducted at the 1.0 mgd Nagog Pond Water Treatment Facility. Secondary treatment for this source includes chlorination, pH adjustment using sodium hydroxide, and fluoridation before the treated water is distributed to the service area. The process is shown in Figure 4.19.

Construction and Operation of the UV Unit

Raw Nagog Pond water was withdrawn from the intake area and flowed through the meter vault on the way to the ozonation facility. The meter reads 0-800 gpm and outputs a 4-20 mA signal to a display and recorder unit in the Nagog Pond facility. The normal flow was expected to be 700 gpm, however, hydraulic headlosses through the 6 inch piping reduced the flow to 300 gallons per minute through the 2002 trial. A booster pump was added at the ozone outlet piping in spring 2003 to increase flow to the target value of 700 gpm.

Raw water was ozonated at Ozone chamber #1 (the south train). Typical ozone doses ranged from 2-4 mg/L, though the system is capable of doses of 6 mg/L at flows of 1 million gallons a day. The existing ozone system PLC monitors ozone dose as lb/MG (pounds per million gallons).

The ozone tank drain at the end chamber of the ozone system was revised such that ozonated water was diverted from the normal outlet to the 6 inch ductile iron pipe. The drain pump was valved off so the ozonated water is diverted to the new vertical tee and pass through the UV unit inlet piping. At this point, the piping changes from 6 inch to 12 inch via a reducer to accommodate the UV unit. After the pipe size increase is a static mixer with injection ports. This is needed for the challenge trial portions of the experiment. The water passes through the static mixer and to a straight length of pipe ahead of the UV unit. The piping has a sample tap so that the concentration of materials or solutions introduced at the static mixer can be sampled ahead of the UV unit. The water was irradiated as it passes through the UV unit. After the UV unit, there is a section of pipe with another sample tap so that the ozone and UV treated water was collected as needed during the experiments. The piping then reduces down to six inch diameter so that the water can enter the bypass piping for ozone train 2 (north train). The water returns the full length of the bypass and enters the last chamber of Ozone train 2 and exits the building toward the Route 2 Pump Station.

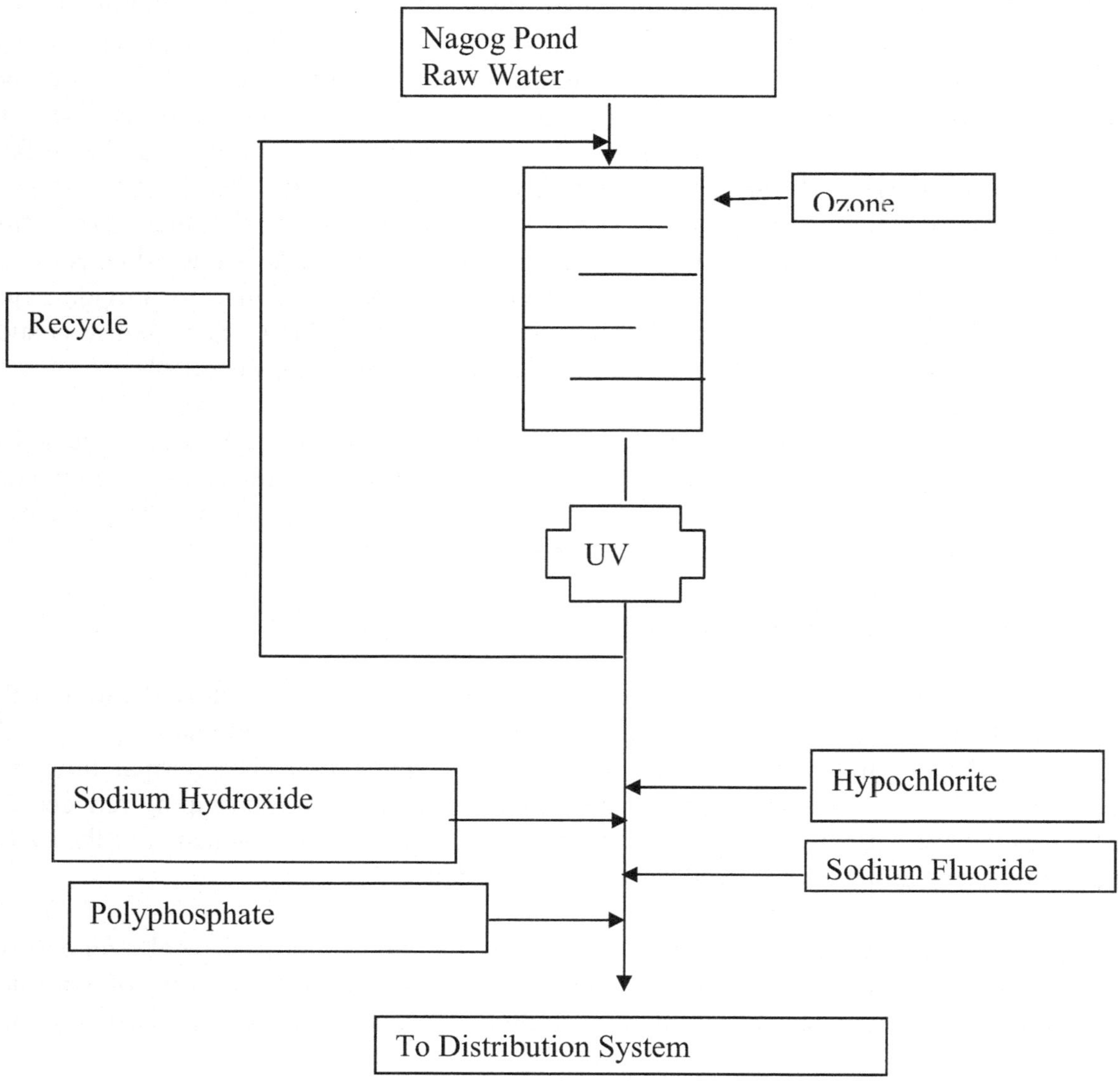

Figure 4.19 Town of Concord Full Scale Pilot at Nagog Pond

A data logger was installed to collect and store information from various instrumentation and control components of the pilot system. The following data were collected continuously:

- Raw water flow – an existing flow meter at Nagog transmits a signal for display (FDT) and to a recorder within the station. The flow signal was split off at the transmitter to send it to data collection.

- Ozone dose – The Hankin ozone unit PLC displays the ozone dosage as mg/L. That signal was sent from the Ozone unit PLC to data collection at the data logger.

- Power usage for the UV system.

- UV dose – the Swift UV unit has two sensors that send 4-20 mA signals of light intensity back to the local control panel (LIT). The UV dose signal was picked up independently and sent to the data logger.

- pH and Temperature – The existing raw water pH and temperature sensor/transmitter that can send a 4-20mA signal to the data collection system was used. The intake at Nagog is shallow enough that there may be some significant daily temperature fluctuations.

- Turbidity – the Nagog Pond Ozone Facility has an existing turbidimeter that records raw water turbidity. These data were collected for the duration of the pilot testing.

A summary of the operational data from the datalogger and from the Route 2 Pump Station records for the first summer of operation of the full scale ozone/UV pilot at Nagog Pond is presented in Table 4.9. The Nagog Pond Facility was on-line from July 5, 2002 to August 31, 2002. The UV unit was on-line from July 5 to August 14, 2002. UV dose was estimated from the voltage readings on the data logger and the relationship between voltage and UV application developed from the March 14, 2002 challenge trial.

Table 4.9
Operational data for the Nagog Pond Ozone Facility during 2002

	Flow (mgd)	Ozone dose (mg/L)	Ozone residual (mg/L)	Estimated UV dose (mJ/cm^2)
Average	0.65	7.17	0.23	162
Minimum	0.46	6.26	0.07	156
Maximum	1.01	12.6	0.49	169

A summary of the operational data from the data logger and from the Route 2 Pump Station records for the second summer of operation of the full scale ozone/UV pilot at Nagog Pond is presented in Table 4.10. The Nagog Pond facility was on-line from July 7 to August 21, 2003. The ozone system was started at noon on July 6 and the UV unit was powered on July 7 when water was sent to the distribution system. The UV unit was set at 100% lamp power. UV dose was estimated from the voltage readings on the data logger and from the UV sensor and the relationship between voltage and UV application developed from the June 3, 2003 challenge trial.

Table 4-10
Operational data for the Nagog Pond Ozone Facility during 2003

	Flow (mgd)	Ozone dose (mg/L)	Ozone residual (mg/L)	Estimated UV dose (mJ/cm^2)
Average	0.80	5.7	0.18	42
Minimum	0.09	0.29	0.10	40
Maximum	0.99	9.08	0.44	46

The addition of a booster pump at the Nagog Pond Facility increased the average flow from Nagog to the system in 2003 as compared to 2002. Required ozone doses to achieve the target ozone residual for 2003 were lower than those required to obtain the desired residual in 2002, indicating better water quality during 2003. The improved water quality may result from a number of precipitation events in the spring and early summer of 2003 which may have diluted natural organic levels in the pond. In contrast, the entire state of Massachusetts was in drought throughout 2002.

The estimated UV doses reflect the aging of the UV lamps over the course of the eighteen months that they were in intermittent service at the Nagog Pond Ozone Facility.

UV Scan Data

While the Nagog Pond Ozone Facility was on-line through the summer of 2002, the University of New Hampshire obtained weekly samples for UV scans. A sampling of the weekly UV scans of raw water, water after ozone application, and water after ozone and UV irradiation are presented in Figures 4.20 and 4.21. The UV scan conducted on the samples is a modification of the SM 5910B Ultraviolet Absorbance Method where the scan varies over the range of ultraviolet wavelengths from 190 nanometers to 400 nanometers, rather than obtaining the absorbance at 254 nanometers as specified in the analytical method. The scan is preferable to the single wavelength data because medium pressure UV radiation occurs at many wavelengths throughout the ultraviolet spectrum and the scan can determine variations in UV absorbance in the water throughout this spectrum.

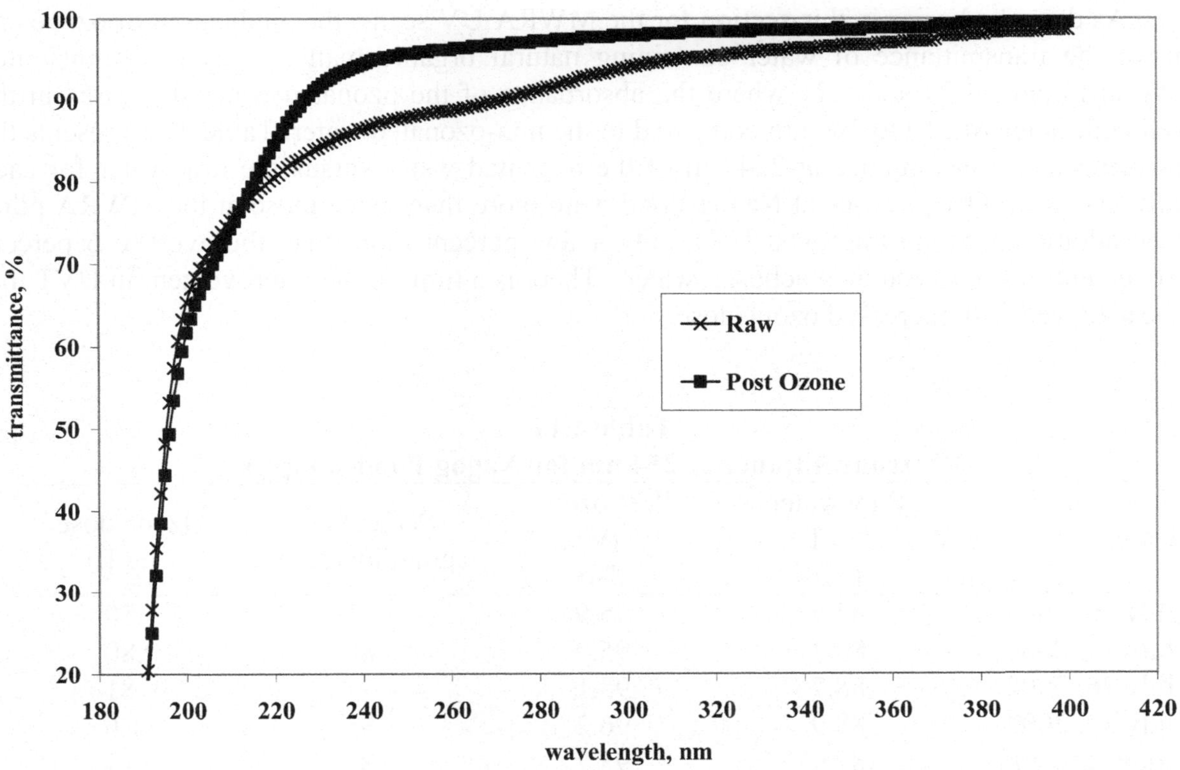

Figure 4.20 UV transmittance scan of Nagog Pond water, July 17, 2002

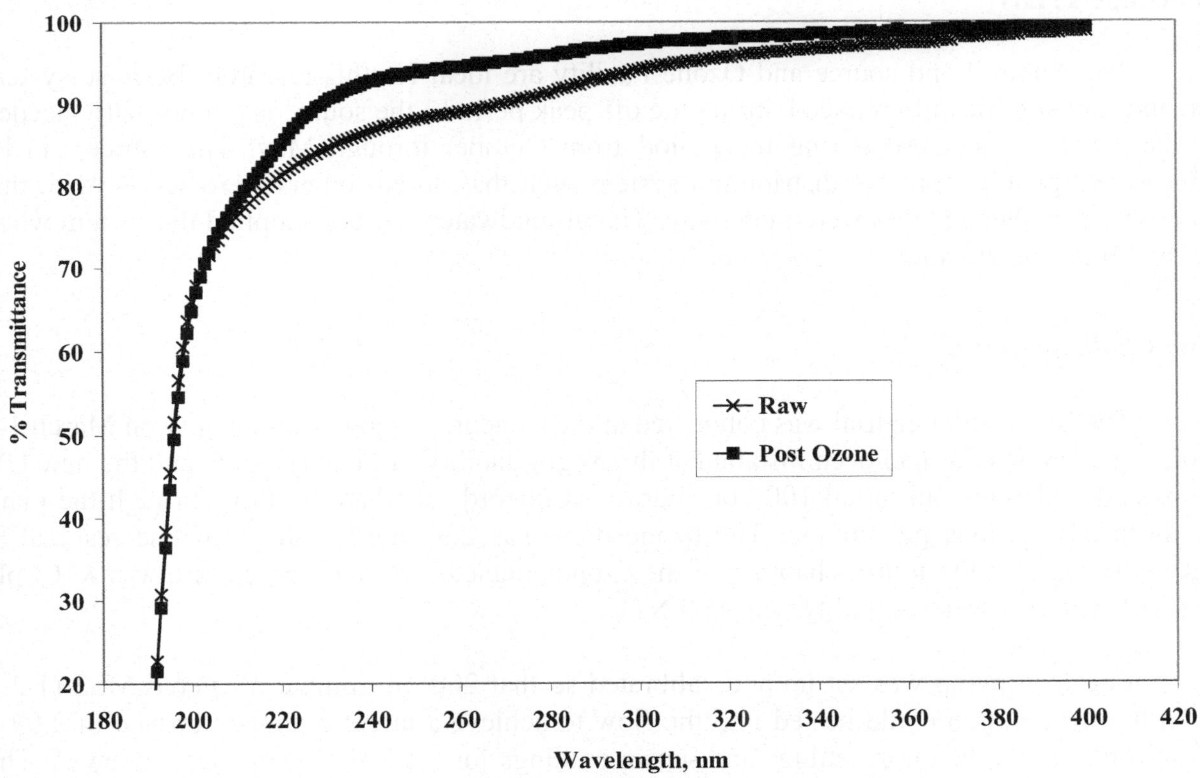

Figure 4.21 UV transmittance scan of Nagog Pond water, June 3, 2003

As described prior in this section for the MWRA UV scans, the application of ozone can improve the transmittance of water containing natural organic matter. This is demonstrated clearly in Figures 4.20 and 4.21 where the absorbance of the ozonated water decreases in the wavelength range of 210 to 340 nm compared to the non-ozonated water. Table 4.11 presents the improvement in transmittance at 254 nm of the ozonated water versus the raw water for each weekly UV scan. Ozone doses at Nagog Pond were more than twice those at the MWRA pilot, but the additional ozone improved UVT only a few percent more than the average 6 percent improvement in UVT seen in Wachusett water. There is a limit to the improvement in UVT that can be achieved with increased ozone dose.

Table 4.11
UV transmittance at 254 nm for Nagog Pond samples

Date	Raw water UVT (%)	Post ozone UVT (%)	Percentage point increase	Ozone dose (mg/L)
July 10, 2002	87.5	95.9	8	7.50
July 17, 2002	88.3	95.9	8	9.80
July 26, 2002	88.7	95.1	7	9.81
July 31, 2002	88.9	96.2	8	11.6
August 5, 2002	87.9	95.7	8	9.17

Challenge Trials

The Nagog Pond source and Ozone Facility are ideal for this research. Because system demands are significantly reduced during the off-peak periods, the source is not normally needed by the Town of Concord during the period from October through May. The source can be isolated completely from the distribution system such that no customers were served with this water while challenge trials were underway. The groundwater sources supplied the system when Nagog Pond was off-line.

First Challenge Trial

The first challenge trial was conducted at the Concord Nagog Pond Facility on March 14, 2002. The UV reactor had been installed at the Nagog facility in January, 2002 and the new UV lamps had undergone an initial 100 hour burn-in. Concord staff had the flow through the plant steady at 310 gallons per minute. The ozone dose was 2.96 mg/L and the ozone residual at 0.20 ± 0.03 mg/L at the fourth chamber of the ozone contactor. Water temperature was 8 °C, pH was 6.65, and raw water turbidity was 0.60 NTU.

The feed pump was set up and calibrated so that 200 mL/minute of titered MS2 (1.2×10^{10} pfu/mL) solution could be fed into the flow to achieve a target organism dosage of 2.09×10^6 pfu/mL. Next, the lamp settings and sensor readings for the UV system were reviewed. The lamp output (4-20mA) was converted to a 1-5 V signal to go to the data logger. The first part of the trial was conducted with ozone feed on, and in accordance with the procedure outlined earlier in this chapter. When the three lamp power setting had been tested, the ozone feed and MS2 feeds were shut off and the sample taps closed. Over the next 45 minutes, ozone readings were taken within the ozone contact basin and at sample taps to determine that the ozone had dissipated from the basin, as the second portion of the trial would be conducted without ozonated water. Once the ozone concentration was below the minimum detectable level, the MS2 feed and sample taps were restarted. Samples were again collected in triplicate as in the first trial and the same three lamp settings were tested. Quality control samples were collected with the UV lamp power off to finish the trial.

The results of the biodosimetry study are presented in Table 4.12. Review of the data shows that the UV doses being delivered by the Trojan Swift unit exceed the anticipated dose range of 40 to 80 mJ/cm^2 and thus completely inactivated all of the MS2 present in the influent to the UV reactor at the 100% and 67.3% power levels. This is primarily due to the fact that the flow to the Swift unit was limited to 310 gpm due to hydraulic constraints whereas the original design flow was planned to be about 694 gpm (1 MGD). The data at 30% lamp power was in the expected range of reduction equivalent doses. Table 4.12 data also show that the UV sensor was responsive to the change in lamp power as would be expected. A small but not statistically significant difference between the preozonated and non ozonated samples was observed, suggesting slightly better reduction equivalent dose at 30 percent lamp power after ozonation.

Table 4.12

Biodosimetry results for the UV reactor testing at Concord, MA on March 14, 2002

Condition	Lamp power level (%)	Mean influent MS2 (log)	Mean effluent MS2 (log)	Mean MS2 inactivation (log)	RED* (mJ/cm^2)
Ozonated	100	6.46	<1.00	>5.89	>110
	67.3	6.46	<1.00	>5.90	>110
	30	6.46	2.07	4.39	73.4
No ozone	100	6.46	<1.00	>5.86	>110
	67.3	6.46	<1.00	>5.86	>110
	30	6.46	2.17	4.29	70.9

* Calculated from the collimated beam data for both ozonated and non-ozonated water using a linear regression analysis with the resulting equation: log MS2 Inactivation = 0.0404 (UV dose) + 1.424; r-squared = 0.991

Second Challenge Trial

The second challenge trial was conducted at the Concord Nagog Pond Facility on June 3, 2003. The UV unit at this time had been in place for 18 months, and the UV lamps were the original units with 1400 hours of operating time. Concord staff maintained the flow through the plant steady at 700 gallons per minute (because of the booster pump addition). The ozone dose was 2.84 mg/L and the ozone residual at 0.20$\pm$0.03 mg/L at the fourth chamber of the ozone contactor. Water temperature was 16.5 °C, pH was 6.16, and raw water turbidity was 0.51 NTU.

The procedure used for this trial is the same as that used for the first trial on March 14, 2002. The stock MS2 solution titered at 6.14 x 10^{11} pfu/mL. The results of the biodosimetry study are presented in Table 4.13.

Table 4.13
Biodosimetry results for the UV reactor testing at Concord, MA on June 3, 2003

Condition	Lamp power level (%)	Mean influent MS2 (log)	Mean effluent MS2 (log)	Mean MS2 inactivation (log)	RED* (mJ/cm^2)
Ozonated	100	6.54	4.24	2.30	43
	67.3	6.54	5.27	1.27	21.5
	30	6.54	5.90	0.64	8.4
No ozone	100	6.32	4.20	2.12	39
	67.3	6.32	5.32	1.00	15.7
	30	6.32	5.70	0.62	8.4
Control	0	6.54	6.50	0.04	N/A

* Calculated from the collimated beam data for both ozonated and non-ozonated water using a linear regression analysis with the resulting equation: log MS2 Inactivation = 0.0482 (UV dose) + 0.2421; r-squared = 0.983

In this trial, it is clear that preozonation benefited the performance of the UV, permitting increased inactivation of the MS2, as compared to the non-ozonated water.

Distribution System Data

Concord has historically sampled monthly at ten locations throughout the distribution system and less frequently at the two storage tanks that serve the Concord system. Six of the ten sample sites are at schools, two sites are fire stations, and the remaining two are public safety and works sites that are staffed at all times. For the full-scale ozone/UV trial, where ozonated and UV irradiated Nagog Pond water was supplied to the distribution system after chlorination, fluoridation and pH adjustment, monitoring was conducted twice per month for most of the water quality parameters, and monthly for disinfection by-products. The sampling protocol is outlined in Table 4.14.

Table 4.14
Distribution system sampling protocol

Parameter	Location	Frequency
Chlorine residual	12 Sites*	Every two weeks
Temperature	10 Sites	Every two weeks
pH	10 Sites	Every two weeks
Turbidity	10 sites	Every two weeks
Iron	10 Sites	Every two weeks
Color	10 sites	Every two weeks
Alkalinity	10 sites	Every two weeks
Hardness	10 sites	Every two weeks
Total coliform	12 sites	Every two weeks
Trihalomethanes	10 Sites	Once a month
Haloacetic acids	10 sites	Once a month
Total organic carbon	5 Sites	Once a month

* 10 regular sites plus two storage tanks

The data generated for the distribution system monitoring sites during the Ozone/UV trial were compiled separately from the data obtained when the Nagog source was not on-line over the course of the two years that Concord ran the ozone/UV trials. Comparison was made between the two sets of data to determine the effect of the addition of the ozone and UV treated surface water source on distribution system water quality.

Table 4.15 presents chlorine residual and temperature data from twelve distribution system sites. There are four locations (most on the western side of the distribution system) that demonstrate the effect of the addition of the Nagog Pond source by differences in chorine residual. Much of the difference in chlorine residual at these sites reveals the change in flow pattern within the distribution system when this significant source is on line. The schools were not instructive in demonstrating differences during typical and trial periods because they are only in use during a non trial time frame of September through June.

Table 4.15
Distribution system temperature and chlorine residual data

Location	Typical		Trial	
	Chlorine residual (mg/L)	Temperature (°C)	Chlorine residual (mg/L)	Temperature (°C)
School 1	0.40	12.5	1.35	20.7
Public 1	0.46	13.9	1.20	21.2
School 2	0.27	13.7	0.23	17.3
School 3	0.47	12.7	0.44	16.7
Public 2	0.27	14.9	0.37	19.6
Public 3	0.50	14.1	0.68	19.9
School 4	0.41	14	0.48	20.8
School 5	0.38	13.3	0.26	19.6
School 6	0.36	15.9	0.38	20.7
Public 4	0.49	14.4	0.56	15.1
Tank 1	---	0.27	0.37	---
Tank 2	---	0.81	0.38	---

The differences in temperatures at the sample sites between normal sampling events and trial sampling events makes sense, as the trial periods occurred during the warm weather months of July and August.

A summary of critical chemical and physical analyses for the ten sampling sites is presented in Table 4.16. The influence of the soft, low alkalinity Nagog Pond source can be observed at two of the schools and one public site.

Table 4.16
Distribution system data for pH, alkalinity, and turbidity

Location	Typical			Trial		
	pH	Alkalinity (mg/L as $CaCO_3$)	Turbidity (NTU)	pH	Alkalinity (mg/L as $CaCO_3$)	Turbidity (NTU)
School 1	7.23	51	0.27	6.86	14	0.69
Public 1	7.21	44	0.43	6.89	14	0.49
School 2	7.23	47	0.27	7.13	24	0.44
School 3	7.17	48	0.22	7.14	42	0.37
Public 2	7.16	48	0.30	7.11	47	0.33
Public 3	7.27	49	0.32	7.18	46	0.24
School 4	7.38	48	0.28	7.24	46	0.22
School 5	7.63	50	0.25	8.09	56	0.21
School 6	7.38	50	0.28	7.56	55	0.17
Public 4	7.41	47	0.25	7.51	56	0.32

No significant differences in color or iron values were observed during the trial compared to the normal background concentrations for these parameters. There were no coliform positive samples at any distribution system site in 2002. One sample had a non-sheen colony, indicating an organism capable of growing on MF media but not a coliform as defined by Standard Method 9222. In 2003, one coliform positive sample was detected at a school site during the ozone/UV trial. The chlorine residual at the time of sampling was less than the minimum detectable concentration.

Heterotrophic plate count data have also been collected periodically at the distribution sample sites. In 2002, seven of the twelve sites demonstrated no difference in heterotrophic plate count numbers during the ozone/UV trial as compared to the non-trial period. At one school, heterotrophic plate count numbers decreased during the trial period. At two public sites and at two schools, heterotrophic plate count numbers increased during the trial period. During 2003, heterotrophic plate counts were obtained only during the trial period. Of the thirty samples collected over four weeks, one sample at a school site exceeded 300 colony forming units per mL. The same site also showed the presence of total coliform and lacked a chlorine residual on the day of sampling.

In both 2002 and 2003, five sets of disinfection by-product samples (trihalomethanes and haloacetic acids) were collected. With the exception of the January sampling event, each set consisted of samples for each of the by-products at the ten sample locations. The January set consisted of four sample locations as required by regulation. Disinfection by-product sampling results are presented in Table 4.17.

Table 4.17
Disinfection by-product data for Concord, MA

Date	THM		HAA	
	Range (mg/L)	Average (mg/L)	Range (mg/L)	Average (mg/L)
January 2002	0.002 – 0.005	0.004	ND*	
March 2002	ND – 0.007	0.004	ND – 0.0023	<0.001
April 2002	0.001 – 0.014	0.006	ND – 0.003	<0.001
July 2002†	0.001 – 0.059	0.019	ND – 0.031	0.009
October 2002	0.002 – 0.018	0.008	ND – 0.007	0.002
January 2003	0.002 – 0.007	0.005	ND*	
April 2003	0.003 – 0.013	0.006	ND*	
July 2003†	0.003 – 0.030	0.012	ND – 0.030	0.006
August 2003†	0.007 – 0.086	0.025	0.002 – 0.033	0.014
October 2003	0.001 – 0.018	0.011	ND – 0.002	<0.001

*None detected
†Nagog Ozone/UV Trial Period

The addition of the Nagog Pond surface water did increase disinfection by-product levels as would be expected in any utility where surface water makes up a proportion of the supply, particularly during warm water and weather conditions. While none of the average disinfection by-product values exceeded the MCLs for trihalomethanes or haloacetic acids, one sample did exceed the trihalomethane standard in August of 2003. The remaining values were well below the trihalomethane limit and no samples approached the haloacetic acid limit of 0.060 mg/L, with the highest haloacetic acid value at 0.033 mg/L. Review of historical data revealed that the sample location that exceeded the 0.080 mg/L trihalomethane value in August 2003 had also exceeded that value in prior years. While the addition of UV to the ozone process at Nagog Pond did not decrease disinfection by-product formation, it did not appear to exacerbate it either.

Two sets of total organic carbon samples were collected at five locations within the distribution system in 2002. Table 4.18 presents these data, which illustrate the mix of surface water with natural organic matter and groundwater with no detectable organic matter at different locations within the system.

Table 4.18
Total organic carbon data from the Concord distribution system

Date	Locations				
	Public 1	Public 2	School 2	School 1	Public 4
8/02/2002	2.55	<1	2.16	2.51	<1
8/22/2002	2.14	<1	1.54	2.27	<1

Overall, addition of the ultraviolet disinfection process to the ozonated surface water did not adversely affect distribution water quality, and it did provide the benefit of significant log inactivation for protozoan and microbial contaminants in addition to the inactivation already provided by ozone and chlorination.

CHAPTER 5
DESIGN AND OPERATIONAL GUIDANCE

This research study has found that combining the strengths of ozone (oxidation, taste and odor reduction, *Giardia* and virus inactivation) and UV (inactivation of *Cryptosporidium* and *Giardia*) can result in a cost-effective, multi-barrier treatment strategy. This chapter will provide two examples of utilities with ozone systems that piloted a UV system to determine the effects of ozone and UV working together. These examples will enable water utilities with ozone or utilities considering installing ozone and UV to find the most cost-effective way of ensuring microbiologically safe water.

BENEFITS AND DETRACTIONS

Ozone

Overall, ozone is an excellent disinfectant for bacteria, viruses, and certain protozoa such as *Giardia*. However, if the target organism is *Cryptosporidium*, the CT values could be 15 to 25 times that of *Giardia*, depending on water temperature (USEPA, 1991; USEPA, 2003a). Also, ozone does not provide a residual to be carried into the distribution system, therefore it can only be used as a primary disinfectant and a secondary disinfectant such as chlorine will be required to maintain a residual in the distribution system.

Ozone typically does not form halogenated disinfection byproducts (TTHMs and HAA5s), but if bromide ion is present in the raw water, then halogenated DBPs may be formed. The addition of free chlorine may actually produce more or less DBPs depending on the nature of the organic material following ozonation. However, free chlorine can be added later in the treatment process after the DBP precursors have been removed and at a lower dose to minimize DBP formation.

Ozone is a strong oxidant, second only to hydroxyl free radical, reducing iron, manganese, color, taste and odor, present in the water. Ozone can react by: direct oxidation of compounds by molecular ozone and/or oxidation of compounds by hydroxyl free radicals produced during the decomposition of ozone. Oxidation of iron and manganese results in converting ferrous (2^+) iron into ferric (3^+) and converting 2^+ manganese to the 4^+ state. The oxidized forms will precipitate as ferric hydroxide and manganese hydroxide.

The ozone system does have a high capital cost because not only does it include a generation system with power supply, but also a gas feed system, an ozone contacting system which can consist of a large concrete basin or pipeline, and a off-gas destruct system. However, ozone does have a relatively low operating cost, ranging from $0.80 to $1.20 per pound of ozone, because the raw materials include only operating power and oxygen (either purchased or generated onsite).

UV

UV is an excellent disinfectant for both *Cryptosporidium* and *Giardia*, regardless of the water temperature, requiring low doses for inactivation. However, UV is not an effective disinfectant for some viruses, such as Adenovirus, requiring doses greater than 100 mJ/cm^2 (includes safety factors) to achieve 2-log inactivation (USEPA, 2003b). But research has indicated that UV can be an effective disinfectant for viruses such as Poliovirus, Hepatitis A, and Rotavirus (Chang et al. 1985, Snicer et al. 1998, Wilson et al. 1992).

UV does not form halogenated DBPs at doses typically used in drinking water treatment (Malley et al. 1996). Therefore systems can use UV to meet disinfection requirements without exceeding DBP MCLs. Also, UV does not have a residual. Therefore, a chemical such as chlorine will have to be added to the finished water to ensure a residual throughout the distribution.

UV alone is not a strong oxidant, but in the presence of hydrogen peroxide (H_2O_2) it can be a very good oxidant through the formation of highly reactive hydroxyl free radicals. However, the doses required to use UV as an oxidant can be greater than 300 mJ/cm^2, compared to a dose of 40 mJ/cm^2 typically used in drinking water disinfection (Linden et al. 2002).

Water quality can have an effect on the performance of the UV system. UV lamps are encased in quartz sleeves, which are subject to fouling in waters with high iron, manganese and/or hardness concentrations. Fouled sleeves reduce the UV light entering the water thereby reducing the ability of the UV system to meet the target dose and requiring an increase in power. Also, waters with a high UV absorbance can require oversized UV reactors or more energy to achieve the required dose, resulting in increased equipment and power costs.

Overall, UV has relatively low capital and operating costs. The equipment required for a UV system consists of a stainless steel reactor with lamps, quartz sleeves, ballasts and control panels. The operating cost of the UV system can range from $0.005 to $0.03 per 1,000 gallons with the raw material being power.

Ozone and UV in Combination

As discussed in the previous paragraphs, ozone and UV have advantages and disadvantages, which are summarized in Table 5.1. However, using both disinfectants in combination at a water treatment plant allows the systems to be optimized, utilizing the benefits of both technologies. Ozone and UV together can provide the oxidation and inactivation required with very low DBP formation and could result in savings in operation costs, especially energy.

Table 5.1
Benefits and detractions of ozone and UV

	Ozone	UV
Bacteria and viruses	Excellent	Good
Protozoa	Good	Excellent
Regulated DBP formation*	Low	Few at required dose
Residual	None	None
Oxidant	Very good	Only with H_2O_2
Microflocculant	Yes	No effect
Capital cost	High	Low
Operating cost	Low	Low

* bromate formation may be a problem in some waters

One of the benefits of utilizing both ozone and UV is that they would provide an excellent multiple-barrier treatment approach. Ozone followed by UV provides a redundant system for *Giardia* inactivation and potentially for *Cryptosporidium* inactivation (under certain conditions). Also, if the ozone system were to go offline, the UV system could meet the inactivation requirements effectively, albeit by brining on a spare reactor or operating at a flow rate lower than design. A strong oxidant like ozone can also increase the UV transmittance of the water, which can result in a smaller, more affordable UV reactor.

The UVDGM recommends that unfiltered systems be allowed to operate under off-specification conditions to produce up to 5 percent of the total monthly volume of water delivered to the distribution system. The design transmittance was selected based on the 95th percentile. This does impose limitations on the treatment operation. Maintaining design transmittance at design flow will become one of the operating parameters for the preceding treatment process, i.e., ozone.

However, both systems in combination would mean an increase in overall cost, capital and operating, especially if a system does not have ozone in place already. Also with both ozone and UV, the overall system is more complex because of the equipment and would require more training and maintenance.

EXAMPLES

MWRA

MWRA has a 405 mgd water treatment facility under construction at Walnut Hill. It is scheduled for completion in 2005. The new Walnut Hill Water Treatment Plant will provide primary disinfection using ozone, corrosion control using sodium carbonate and carbon dioxide, secondary disinfection with chloramines, and fluoridation from application of hydrofluosilicic acid. Ozone design parameters for the new Walnut Hill Water Treatment Plant are outlined in Table 5.2

Table 5.2
Walnut Hill ozone design parameters

	Average	Minimum	Maximum
Flow, mgd	220	150	405
Ozone dose, mg/L	2.0	1.0	6.0
Contact time, min	30	20	60

The pilot scale work described in Chapter 4 presents information from pilot scale testing of UV disinfection alone and ozone followed by UV disinfection as primary disinfection schemes for Wachusett water. In addition to providing MWRA operations staff with hands on experience using medium pressure UV system, the ultraviolet system demonstrated log reduction of MS-2 on the order of 3.2 with new lamps to 1.9 with aged lamps.

The use of ozone prior to UV disinfection resulted in treated water with lower halogenated disinfection by-products and improved clarity and UV transmittance of the water. The ozone did create substantial amounts of aldehydes and keto-acids, both of which are bio-degradable and neither of which are as amenable to formation of halogenated by-products as less oxygenated compounds. Table 4.3 presented a summary of the UV transmittance of the water with and without ozone. The data were used to develop a cumulative frequency percentile chart indicating the percentage of time that the data were lower than a given value of UV transmittance. The cumulative frequency chart is presented as Figure 5.1. The 95th percentile UV transmittance value of the ozonated water was 93 while the 95th percentile value of the non-ozonated water was 85.

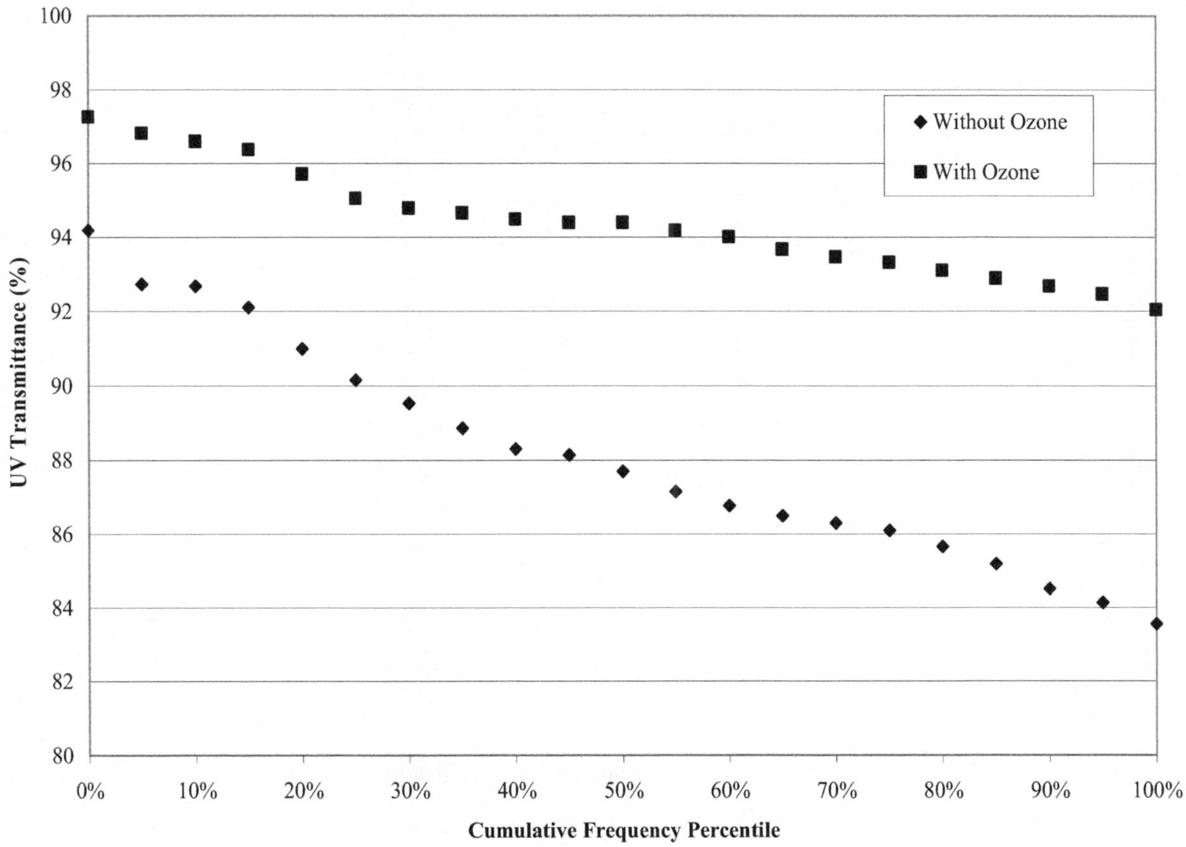

Figure 5.1 Ozone Effects on UV transmittance of Wachusett water from October 2002 to October 2003

The transmittance data from this chart were used as the basis for the conceptual design of two different UV disinfection options, the first with UV applied to the raw water ahead of the Walnut Hill Water Treatment Plant based on a design transmittance of 85 percent and the second after ozone application with a design transmittance of 93 percent. It should be noted that during the conceptual designs, the design UV transmittance values were based on data from October 2002 to July 2003, with the 95th percentile UVT for the without ozone option was 85 percent. With the addition of August 2003 to October 2003 UV transmittance data, that value changed to 84 percent; the ozone UVT value did not change.

For the second option, it was necessary to estimate the ozone dose required to increase the UV transmittance to 93 percent. Based on data from the pilot study conducted from October 2002 to October 2003, which included raw water UV transmittance, ozone dose and ozonated water UV transmittance, the minimum ozone dose required to increase the average raw water transmittance from 89.5 percent to 93 percent was calculated. The ozone dose was determined using a multiple regression model, which provided a prediction equation, shown in Figure 5.2. Based on the prediction equation, a 1.1 mg/L dose of ozone is required to increase the UVT from 89.5 percent to 93 percent.

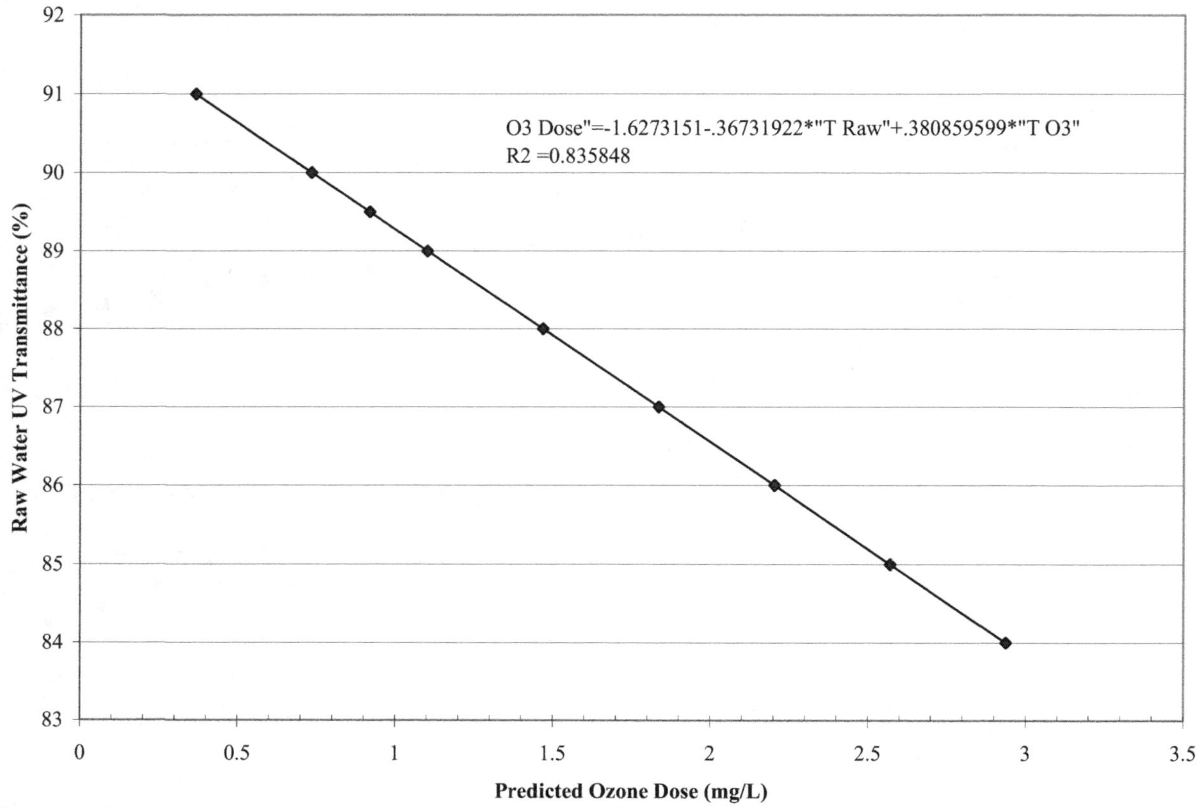

Figure 5.2 Wachusett Water ozone dose prediction equation to achieve 93 percent transmittance

Cost estimates were developed for both options based on flow criteria presented for the Walnut Hill Water Treatment Plant, the ozone facility that is currently under construction for MWRA, and design criteria presented below (Table 5.3). The UV construction cost includes the following equipment and allowances:

- UV reactors (duty and standby), MP or LPHO
- UV intensity monitors – one per lamp for MP or per row for LPHO
- Online transmittance monitor
- Cleaning system – mechanical and/or chemical
- Collimated beam apparatus
- Reference sensors
- Installation
- Building and associated piping and valves

Construction costs for the ozone facility are not included because the ozone facility for MWRA had already been designed and was under construction before this study took place.

The cost also includes allowances for electrical and instrumentation (10% of construction cost), sitework (15% of construction cost), HVAC and plumbing (10% of construction cost), engineering (30% of subtotal), and contingencies (20% of subtotal). The capital cost for the facility does not include piping modifications, modifications to existing facilities, and emergency or backup power costs.

Table 5.3
MWRA UV design criteria and costs

	Option A Pre-ozone UV	Option B Post-ozone UV
Design UV transmittance, %	85	93
Cryptosporidium inactivation	2 log	2 log
Maximum headloss, inches water	36	36
Water temperature range, °C	1 – 22	1 – 22
Maximum color, PCU	20	10
Capital cost, minimum	\$14,000,000	\$11,000,000
Capital cost, maximum	\$36,000,000	\$29,000,000

Placing the UV after ozone addition resulted in UV capital costs savings of 19 to 21 percent compared to UV as a stand alone technology for this water source. The improvement in UVT afforded by ozone application results in UV units that require fewer lamps to achieve the target inactivation of *Cryptosporidium* required.

Estimates were also made for annual operating and maintenance costs for Options A and B, presented in Table 5.4, assuming an average daily flow of 220 mgd. UV operating costs include energy costs for the UV units and energy associated with implementing the UV addition. Maintenance costs include labor for cleaning and maintaining the UV units and equipment replacement costs. Overall, the UV O&M costs for Option B were lower than the UV O&M costs for Option A. The smaller amount of lamps required for Option B translated to savings in energy and maintenance. However, the ozone O&M costs must be taken into consideration also for Option B. The ozone O&M costs were estimated at \$580,000 per year based on an ozone dose of 1.1 mg/L (to improve the average water transmittance from 89.5 percent to 93 percent), power cost of \$0.10 per kWh, generator power of 4.5 kWh per pound of ozone, ozone concentration of 12 percent, and liquid oxygen cost of \$80 per ton of oxygen.

Table 5.4
MWRA Walnut Hill annual operating cost estimates

	Option A Pre-ozone UV	Option B Post-ozone UV	Option B Total (ozone and UV)
Minimum cost	\$700,000	\$500,000	\$1,080,000
Maximum cost	\$2,500,000	\$1,700,000	\$2,280,000

Table 5.4 shows the operating cost range for UV and ozone at MWRA. In both cases (minimum and maximum operating costs), the operating costs for the post-ozone UV system were lower than the pre-ozone system. Even when taking the ozone system operating cost into account, the maximum operating cost is lower for the ozone/UV system. The minimum operating cost (associated with the bottom of the estimated range) shows that the ozone/UV system would cost

more to operate than the UV alone. In that case, there are still other benefits to operating both systems.

The present worth costs for the two options were also determined, as shown in Table 5.5. A 20-year present worth analysis was conducted using a discount rate of 6.0 percent and includes capital costs, UV O&M for Option A and UV and ozone O&M for Option B.

Table 5.5
Present worth analysis

	UV Capital Cost	Operating cost	Total present worth cost
Option A, minimum cost	$14,000,000	$700,000	$22,029,000
Option A, maximum cost	$36,000,000	$2,500,000	$64,675,000
Option B, minimum cost	$11,000,000	$1,080,000	$23,387,600
Option B, maximum cost	$29,000,000	$2,280,000	$55,151,600

Based on this analysis, Option B, ozone followed by UV disinfection, had the lower total present worth cost. Therefore, the improvement in UV transmittance provided by the ozone system decreased not only the capital cost for the UV system but also the total present worth of the system. However, for a system that does not have ozone already in place, the addition of an ozone system to improve UV transmittance would not be cost effective because of the additional capital cost.

The addition of UV, before or after ozone application, would permit MWRA to apply an ozone dose that may not necessarily provide 2 log *Cryptosporidium* inactivation as that role would be fulfilled by the UV units. Thus, the ozone dose applied to Wachusett water could be determined by other water quality considerations such as virus inactivation, taste and odor removal, or UVT improvement if the UV were placed after the ozone. MWRA should evaluate ozone operation to determine scheduled maintenance down time as well as unanticipated down time to see if this time is less than the 5% off-specification allotment.

CONCORD, MA

Concord, Massachusetts operates the 1 mgd Nagog Pond Ozone Facility which provides primary disinfection to unfiltered Nagog Pond water before the water flows to the Route 2 Pump Station for secondary disinfection, post-treatment and distribution. The Town of Concord is considering the installation of a UV system for several reasons. One reason is that UV disinfection is very effective for inactivation of *Cryptosporidium* and *Giardia* regardless of water temperature, while the Nagog Pond facility's existing ozone system can be effective for inactivation of these organisms only at a limited range of water temperatures. Also, the ozone system is aging and can not be rehabilitated. UV disinfection alone would satisfy the Long Term 2 Enhanced Surface Water Treatment rule (LT2ESWTR) requirements for 2-log inactivation of *Cryptosporidium*. Therefore, if the ozone system were taken off line, Concord would still have two disinfectants, UV for inactivation of *Cryptosporidium* and *Giardia* and chlorine for inactivation of viruses, which satisfies the requirements of LT2ESWTR.

Design conditions for the Concord UV system were developed from raw water quality data, data from the AwwaRF study, and other Nagog Pond facility design criteria. A summary of the design criteria is presented in Table 5.6. Town of Concord officials have indicated that they anticipate placement of the Nagog Pond facility into Bin 1 (from the LT2ESWTR) which requires 2 log inactivation credit for *Cryptosporidium*. Therefore, the design doses of the UV reactors proposed for the conceptual design are based on 2 log inactivation according to the USEPA UV Disinfection Guidance Manual.

Table 5.6
Nagog Pond WTP UV design criteria

Criteria	Units	Option A Raw Water	Option B Ozonated Water
Peak hourly flow rate	gpm	800	800
Average daily flow rate	gpm	500	500
Minimum flow rate	gpm	300	300
Design UV transmittance	%	87	95
Cryptosporidium log inactivation per UVDGM	log	2	2
Guaranteed maximum headloss through reactor	inches of water column	120	120
Turbidity	NTU	0.2 – 1.2	0.2 – 1.2
Maximum hardness	mg/L as $CaCO_3$	24	24
Maximum iron concentration	mg/L as Fe	0.6	0.6
Maximum manganese concentration	mg/L	0.64	0.64
Total organic carbon range	mg/L	1.6 – 9.1	1.6 – 9.1
Maximum color	PCU	25	5
Water temperature range	°C	1 - 26	1 – 26

As shown in Table 5.6, there is a difference between the UV transmittance (UVT) values for raw water and ozonated water, for the two month span that Nagog Pond was on-line in 2002. UVT is a key factor in a UV system design, because it can have an effect on the capacity of the UV reactors and the number of reactors required. The design UVT values in Table 5.5 were based on the 95th percentile. Table 5.7 summarizes the UVT data at 254 nm collected in July and August, 2002 for both raw and ozonated water. Based on these data, ozonation improved the raw water UVT by 8 percentage points.

Table 5.7
Effect of ozone on UV transmittance at Nagog Pond (measured at 254 nm)

Date	Raw water transmittance (%)	Ozonated water transmittance (%)	Percentage point increase
July 10, 2002	87.5	95.5	8
July 17, 2002	88.3	95.9	8
July 26, 2002	88.7	95.1	7
July 31, 2002	88.9	96.2	8
August 5, 2002	87.9	95.7	8

The data from this evaluation were used to determine the design UV transmittance values. A cumulative frequency (CF) diagram was plotted for raw and ozonated water as Figure 5.3. The CF diagram shows the percentages of the dataset that are higher for a given value of transmittance over the collection period. For example, if the 90th percentile UVT is 92 percent, then 90 percent of the measurements were higher and only 10 percent of the measurements were lower than 92 percent over this period.

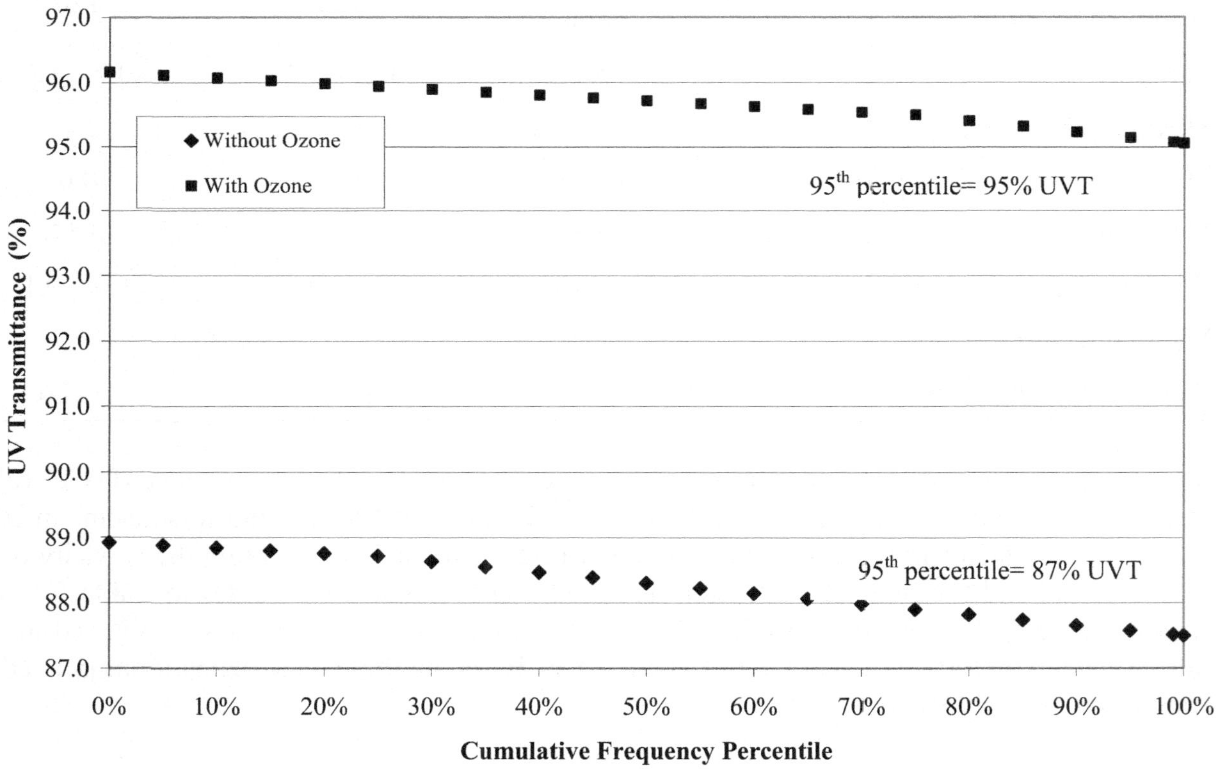

Figure 5.3 Ozone effects on UV transmittance of Nagog Pond from July to August 2002

Given that the Nagog Pond Facility had successfully been re-piped to accommodate the UV unit for the pilot study, full scale designs were developed placing new ultraviolet units and accessories in the existing pipe gallery between the two ozone contact chambers. The construction cost includes UV equipment, allowances for piping, electrical work, improved HVAC and plumbing and contingencies. The cost of medium pressure UV equipment ranged from $100,000 to $160,000 so the median cost of $120,000 was used to prepare the opinion of probable cost table (Table 5.8). The size of the Concord facility and water demands are such that no substantial differences were proposed by UV manufacturers for units treating raw water versus ozonated water even though there is a significant improvement in UVT of the ozonated water.

The probable operations and maintenance cost, assuming an average flow of 500 gpm, is $0.03/1,000 gallons for the LPHO UV reactor and $0.03-0.06/1,000 gallons for the MP UV reactor options. Historically, the Town of Concord has operated the Nagog facility for 90 days a year, but if Concord were to use Nagog every day, annual costs for UV operation would range from $8,000 to $16,000.

Table 5.8
Opinion of probable cost – ozone contactor gallery
Nagog Pond Ozone Facility

Items	Low pressure high output reactor	Medium pressure reactor
Construction cost	$220,000	$250,000
Engineering, legal and administration cost	$ 50,000	$50,000
Total cost	$270,000	$300,000

OPERATIONAL CONSIDERATIONS

The operational staff at MWRA and Concord, MA found the UV systems to be reliable and easy to operate. The internal UV lamp sleeve wipers were set to wipe every four hours that the unit was in operation and the sleeves required cleaning once during the one year pilot. The operators praised the simplicity, low maintenance, and ease of operation (on a macroscopic scope) of the units. The operators would have liked a friendlier user interface and easier access to the unit. The UV pilot units, which were prototypes and did not have all the capabilities of later production models, had some software limitations. Overall, the experience was good and the operations staffs believed that the newer units would be easier to use, maintain, and repair if and when necessary.

FINDINGS

Using both ozonation and UV in a combined disinfection scenario allows both disinfectants to work synergistically. Some of these benefits are listed below:

- In cold water, very high ozone doses and CT values are required to disinfect *Cryptosporidium* cysts. Ozone systems would require extremely long detention times and high ozone residuals concentrations to provide the necessary CT at water temperatures below 5 °C. UV disinfection on the other hand, is insensitive to water temperature.

- Ozone is effective at disinfecting viruses while much higher UV doses would be required based on Adenovirus data in the UVDGM.

- Ozone can provide collateral water quality benefits such as taste and odor elimination, oxidation of dissolved iron and manganese, and destruction of disinfection by-product precursors.

- Ozonation increases a water's UV transmittance, thereby making UV disinfection more effective.

- At high doses, ozone can increase the formation of biodegradable organic matter such as aldehydes and keto-acids, or bromate. By using lower ozone doses in a combined Ozone/UV disinfection scheme, formation of these by-products can be reduced.

- From a design perspective, looking strictly at the capital cost of the UV equipment, a post-ozone UV facility is estimated to be about 20% less expensive than a pre-ozone or no ozone UV facility. This benefit accrues primarily to large systems where substantial savings in numbers of UV units or lamps is seen as the transmittance of the water improves.

- Improved operational controls for UV units may permit utilities to operate ozone and UV more beneficially. The pilot UV units were limited to set increments of lamp power, often providing more UV light and using more electrical power than was necessary to achieve 2 to 3 log *Cryptosporidium* inactivation. More operational control of UV doses with a minimum lamp power at a setting that will ensure the minimum required inactivation would be a tool to save operational costs.

There are some potential disadvantages of Post-Ozone UV Systems:

- When UV reactors are placed after ozonation, care must be taken to ensure that ozone residual is minimized prior to UV. Excessive ozone residual will be a source of ultraviolet absorbance and lower the disinfection effectiveness of the UV.

- If, because of power fluctuations, the UV lamps should go off, the ozone dosage could be increased to inactivate *Cryptosporidium*. However, there will be some water that passes to the distribution system before this adjustment occurs. Thus, off-specification water will pass into the clearwell.

- Fouling by inorganic iron and manganese, if present in the raw water, may be increased by oxidation of both of these metals by ozone ahead of UV disinfection.

RECOMMENDATIONS

Systems that already have ozone can benefit greatly by including the increase in transmittance into the design of the UV system, making it smaller and more energy efficient. In general, systems that don't have ozone and are considering UV would not find the savings in UV system costs to be enough to pay for the ozone system. If ozone is used for other purposes besides increasing transmittance (a second line of defense in a multi-barrier approach to pathogen control, taste and odor control, oxidation, filtration enhancement, color removal, THM/HAA reduction), the utility can see a benefit by installing both technologies.

A utility considering installation of UV following an existing ozone system should evaluate the scheduled and unscheduled down time during the year. Is the amount of time acceptable and could it be counted during the 5% off-specification allotment. Can maintenance of the ozone system be scheduled for those times when the UVT of the water is at its best? The UV design should accommodate the full range of flows with the consideration that flow could be reduced when the ozone system is off-line.

For both existing and new ozone systems, the operational strategy will need to be evaluated. Two potential strategies include using the ozone system to provide some level of disinfection with UV as a second disinfectant or using the ozone to accomplish other water quality goals such as taste and odor removal and UVT improvement with UV as the primary disinfectant.

For smaller systems, the improvement in UVT with ozone application may not translate to lower capital costs as there are limits to design flexibility for both systems. However, the operational and regulatory compliance benefits of ozone and UV outlined above can be seen by small systems.

REFERENCES

APHA, AWWA, and WEF (American Public Health Association, American Water Works Association and Water Environment Federation) 1995. *Standard Methods for the Examination of Water and Wastewater*. 19th ed. Washington, D.C: APHA.

Arrowood M.J., L.T. Xie, K. Rieger, J. Dunn. 1996. Disinfection of *Cryptosporidium parvum* oocysts by Pulsed Light Treatment Evaluated in an *In Vitro* Cultivation Model, *Jour. Eukar. Microbiol.*, 43:(5):S88-S88.

Baron, J. 1997. Repair of wastewater microorganisms after ultraviolet disinfection under semi-natural conditions *Wat. Environ, Res.*, 69 (5): 992-998.

Battigelli, D.A., M.D. Sobsey, and D.C. Lobe. 1993. The Inactivation of Hepatitis A Virus and Other Model Viruses by UV Irradiation. *Water Sci and Tech*, 27(3-4): 339-342.

Black & Veatch. 1995. *Ultraviolet Radiation for the Disinfection of Water and Wastewater*. Report for EPRI-CEC.

Blatchley, E.R., W.L. Wood, and P. Schuerch. 1995. UV Pilot Testing: Intensity Distributions and Hydrodynamics. *Journal of Environmental Engineering,* 121(3):258-262.

Blatchley, E.R., 1997, Numerical Modeling of UV Intensity: Application to Collimated Beam Reactors and Continuous Flow Systems. *Water Research*, 31:2205-2218.

Block, J.C., I. Sibille, and L. Mathieu 1993. Biofilm Accumulation in Drinking Water Distribution Systems. *Biofouling,* 6:333-343.

Bolton, J.R. 1998. Detailed Procedures for the Calculation of the Average Irradiance on an Ultraviolet Disinfection Reactor Using Broadband Medium-Pressure Mercury UV Lamps. Submitted July 1998 to *Water Research*.

Bolton, J.R., and K.G. Linden. 2002. Standardization of Methods for Fluence (UV Dose) Determination in Bench-scale UV Experiments. *J. Environ. Engineering*. (in press)

Bosch, A., C. Tartera, R. Gajardo, J. M. Diez, and J. Jofre. 1989. Comparative resistance of bacteriophages active against *Bacteroides fragilis* to inactivation by chlorination or ultraviolet radiation. *Water Sci. Technol.* 21:221-226.

Braunstein, J.L., J.L. Darby, F.J. Loge, G. Tchobanoglous 1996. Ultraviolet Disinfection of Filtered Activated Sludge Effluent for Reuse Applications, *Water Environment Research*, 68(2):152-161.

Budowsky, E.I., E.D. Sverdlov, and T.N. Spasokukotskaya, 1992. Direct and Indirect Action on Thymine in Aqueous Solution by Powerful Laser Ultraviolet Radiation. *Photobiochem. Photobiophy*, 4: 233-240.

Bukhari, Z. T.M Hargy, J.R. Bolton, B. Dussert, and J.L. Clancy, 1998. Inactivation of *Cryptosporidium parvum* by Medium-Pressure Ultraviolet Light in Finished Drinking Water. In *Proceedings AWWA Annual Conference and Exposition*. Denver, CO.: AWWA.

Bukhari, Z., T.M. Hargy, J.R. Bolton, B. Dussert and J.L. Clancy. 1999. Medium-Pressure UV Light for oocyst inactivation. *J. AWWA*, 91(3)86-94.

Bukhari, Z., M.M. Marshall, D.G. Korich, J.S. Rosen, C.R. Fricker, H.V. Smith, and J.L. Clancy. 2000. Cryptosporidium parvum viability and infectivity: inter-laboratory comparisons to determine correlation between assays. *J. Appl. Environ. Microbiol.* 66:2972-2980.

Campbell, A. T., Robertson, L. J., and Smith, H. V. 1995. Inactivation of Oocysts of *Cryptosporidium parvum* by Ultraviolet Irradiation. *Wat. Res.*, 29: 2583.

Carlson, D.A., R.W.Seabloom, F.B.DeWalle, T.F.Wetzler, J.Engeset, R.Butler, S.Wangsuphachart, and S.Wang. 1985. Ultraviolet disinfection of water for small water supplies. EPA/600/S2-85/092. US Environmental Protection Agency, Washington, D.C.

Chang, J.C.H., S.F. Osoff, D.C. Lobe, M.H. Dorfman, C.M. Dumais, R.G. Qualls and J.D. Johnson. 1985. UV inactivation of pathogenic and indicator microorganisms. *Appl. and Environ. Microbiol,.* 49(6):1361-1365.

Chrtek, S. and W. Popp. 1991. UV disinfection of secondary effluents from sewage treatment plants. *Wat. Sci. Tech*, 24: 343-346.

Clancy J. L. and C. R. Fricker. 1998. Control of *Cryptosporidium* – How Effective is Drinking Water Treatment? In *Proceedings of the WQI Conference*, Vancouver, Canada.

Clancy, J.L., T.M. Hargy, M.M. Marshall, and J.E. Dyksen. 1998a. *Innovative Electrotechnologies for Inactivation of Cryptosporidium parvum.* Denver, CO.: AwwaRF and AWWA.

Clancy, J.,T.M. Hargy, J.E. Dyksen, 1998b. UV light inactivation of *Cryptosporidium* oocysts. *Journal AWWA*, 90(9)92.

Clancy, J.L., Z. Bukhari, J. Bolton, and B. Dussert . 1998c. Inactivation of *Cryptosporidium parvum* Oocysts Using Medium Pressure Ultraviolet Light in Bench and Pilot Scale Studies. In *Proceedings AWWA Water Quality Technology Conference*, San Diego, Cal., Nov 1-4.

Clancy, J.L., M.M. Marshall, C.R. Fricker, and H. Smith, 1998d, Comparison of Viability and Infecitvity Methods for *Cryptosporidium* in the US and UK, In *Proceedings AWWA Water Quality Technology Conference*, San Diego, Cal., Nov 1-4.

Clancy J. L. and C. R. Fricker. 1999. *Giardia* and *Cryptosporidium* Methods for Water Testing - Current Status and Future Needs. *Water Supply*,17 (2): 49-55.

Clancy, J.L., Z. Bukhari, T.M. Hargy, J.R. Bolton, B. Dussert, and M.M. Marshall. 2000. Using UV to inactivate Cryptosporidium. *J. Amer. Water Works Assoc.* 92(9):97-104.

Clancy, J. L., M. M. Marshall, T. M. Hargy, D. Korich, and D.A Battigelli. 2002. Susceptibility of five strains of *Cryptosporidium parvum* oocysts to ultraviolet light disinfection *Appl. Environ. Microbiol.*(in press).

Clement, J., M. Hayes, P. Sarin, W.M. Kriven, J. Bebee, K. Jim, M Beckett, V.L. Snoeyink, G.P. Kirmeyer and G Pierson, 2002, *Development of Red Water Control Strategies*, Denver, Colo, AwwaRF and AWWA

Coffey, Bradley M. 1994.*Demonstration-Scale Evaluation of Ozone Disinfection Calculation Methods*. Presented at the IOA PAG Meeting. Richmond, VA.

Coffey, B.M., Huck, P.M., Maurizio, D.D., Emelko, M.B., Douglas, I.P., and Van Den Oever, J. 1999. "The Effect of Optimizing Coagulation on the Removal of Cryptosporidium parvum and Bacillus subtilis." In: Proc. 1999 AWWA Water Quality Technology Conference, AWWA. Tampa, Florida. Oct. 31 - Nov. 3.

Craggs, J., S.H. Kirk, S.I. Ahmad. 1994. Synergistic Action of Near-UV and Phenalanine, Tyrosine or Tryptophan on the Inactivation of Phage T7: Role of Superoxide Radicals and Hydrogen Peroxide. *J. Photochem. Photobiol. B:Biol*, 24: 123 129.

Darby, J., M.Heath, J. Jacangelo, F. Loge, Swaim, P., and G. Tchobanoglous. 1995. Comparative Efficiencies of Chlorination/Dechlorination and Ultraviolet Irradiation. WERF Report 91-WWD-1

DeMers, L. and R.C. Renner. 1992. In *Alternative Disinfection Technologies for Small Drinking* Water *Systems*, Denver, CO: AWWA Publishing.

Driedger, A.M., J.L. Rennecker, and B.J. Marinas. 1999. Optimization of *C. parvum* Inactivation with Ozone/Free Chlorine and Ozone/Monochloramine. In: *Proc. 1999 AWWA Water Quality Technology Conference*, AWWA. Tampa, Florida.

Finch,G.R.and E.K.Black. 1993. Inactivation of *Giardia* and *Cryptosporidium* using ozone. In *Proceedings of the 11th Ozone World Congress*, San Francisco, CA.

Finch, G.R., E.K. Black, and L.L. Gyurek. 1994. Ozone and Chlorine Inactivation of *Cryptosporidium*. In *Proceedings AWWA Water Quality Technology Conference*. San Francisco, CA.

Finch, G.R., L.L. Gyurek, L.R.J. Liyanage, and M. Belosevic. 1997. Effects of various disinfection methods on the inactivation of Cryptosporidium. AwwaRF, AWWA.

Finch, G.R., N. Neumann, L.L. Gyurek, J.Bradbury, L. Liyanage, and M. Belosevic. 1998. Sequential Chemical Disinfection for the Control of *Giardia* and *Cryptosporidium* in Drinking Water. In *Proceedings AWWA Water Quality Technology Conference*. San Diego, CA.

Finch, G.L. and M. Belosevic. 1999. Inactivation of *C. parvum* and *G. muris* with Medium Pressure Ultraviolet Radiation. Proceedings USEPA UV Workshop. Washington, D.C., USEPA.

Finch G. R. and M. Belosevic. 2000. Inactivation of *Cryptosporidium parvum* and *Giardia muris* with Medium Pressure Ultraviolet Radiation. In Press – Wat. Res.

Finch, G.R., C.N. Haas, J.A Oppenheimer, G. Gordon, R.R. Trussel. 2001. Design Criteria for Inactivation of *Cryptosporidium* by Ozone in Drinking Water. *Ozone Science and Engineering*. 23:259-284.

Gates, F.L. 1929. A Study of the Bactericidal Action of Ultra Violet Light, I. The Reaction to Monochromatic Radiation. *Journal of General Physiology*, (13):231.

Gerba, C., M.N. Karpiscak, L.R. Sanchez, R.J. Freitas, 2001, Enteroviruses: basic biology and disease In: Wastewater Reclamation and Reuse. T. Asano, ed. pp. 437-488. Technomic, Lancester, PA

Gigliotti, T.M., S.J. States and L.W. Casson. 1999. Pilot-Scale Studies of *Cryptosporidium* Inactivation Using Ozone. In *Proceedings AWWA Water Quality Technology Conference*. Tampa, Fla., Oct 31-Nov 3

Gyurek, L.L., H. Li, M. Belosevic and G.R. Finch. 1999. Ozone Inactivation Kinetics of *Cryptosporidium* in Phosphate Buffer. *Journal of Environmental Engineering,* 125 (10):913-924.

Halliday, D. and R. Resnick. 1988. *Fundamentals of Physics*. 3rd ed. NY: John Wiley & Sons.

Hargy, T.M. 1999. The Use of Ultraviolet Light for Inactivation of *Cryptosporidium* in Water. In *Proceedings of the Water Quality Association Annual Conference*. Chicago, Mich., June 20-24

Hargy, T.M. 2000. UV equipment proven against *Cryptosporidium*. Water Conditioning and Purification. March, pp 36-38.

Harm, W. 1980. *Biological Effects of Ultraviolet Radiation*, Cambridge University Press, New York.

Harris, G.D., V.D. Adams, D.L. Sorensen and M.S. Curtis. 1987a.Ultraviolet inactivation of selected bacteria and virus with photoreactivation of the bacteria. *Wat. Res.*, 21(6):687-692

Harris, G.D., 1987b. Potassium Ferrioxalate as Chemical Actinometer in Ultraviolet Reactors. *Journal of Environmental Engineering*, 113:612-627.

Havelaar, A.H. 1991. F-Specific RNA Bacteriophages as Model Viruses in UV Disinfection of Wastewater. *Water Science and Technology*, 24(2):347-352.

Herbold, K., B. Flehming, and K. Botzenhart. 1989. Comparison of zone inactivation, in flowing water, of hepatitis A virus, poliovirus 1, and indicator organisms. *Applied and Environmental Microbiology*, 55 (11): 2949-2953.

Hirata, T. 1999. Effect of Ozonation and Chlorination on Viability and Infectivity of *Cryptosporidium parvum* Oocysts. In *IAWQ/IWSA International Conference Abstract Proceedings.*

Hoyer, O. H. Schumann, C. Von Sontag, 1992. UV Fluence Rate Determination of the Low-Pressure Mercury Arc in the UV Disinfection of Drinking Water. *Aqua*. 41(2):75-81.

Hulsey, R.A., D. Neden, I. Singh, J.H. Clark, S.N. Foellmi and M. Friedman. 1997. Ozone Demand and Decay Studies for an Unfiltered Supply. In *Proc. 1997 AWWA Water Quality Technology Conference*. Denver, CO.

Hulsey, R.A., G.S. Logsdon, H.S. Shorney, and B.W. Long. 1999. Ozone Residuals -- What to Do at the End of the Contactor. In *Proc. 1999 AWWA Water Quality Technology Conference*. Tampa, Florida

Hulsey R., D. McCarthy and C. Lewis. 2000. Ozone counters *Cryptosporidium* in Milwaukee. *World Water and Environmental Engineering*, Jan / Feb 2000.

Jacob, S.M. and J.S. Dranoff. 1970. Light Intensity Profiles in a Perfectly Mixed Photoreactor. *Jour. A.I.Ch,* 16:359-363.

Jagger, J. 1958. Photoreactivation. *Bact. Rev*., 22: 99-142.

Jagger, J.H. 1967. *Introduction to Research in UV Photobiology*. Prentice-Hall, Inc. Englewood Cliffs, NJ

Kanjo, Y., M. Kimata, S. Miyanaga, H. Okada, C. Banno, M. Matsumoto, Y. Shimada. 2000. Inactivation of *Cryptosporidium* spp. Oocysts with Ozone and Ultraviolet Irradiation Evaluated by *In vitro* Excystation and Animal Infectivity. *Water Science and Technology*. 41 (7): 199-125.

Karanis, P., W.A. Maier, H.M. Seitz and D. Schoenen. 1992. UV sensitivity of protozoan parasites. *Aqua*, 41:95-100.

Kitchens, J.F., Casner, R.E., Edwards, G.S., Harward, W.E., and Macri, B.J., 1976 Investigation of Selected Potential Environmental Contaminants: Formaldehyde, USEPA ARC-49-5681, Washington D.C.

Korich, D.G., M.M. Marshall, H.V. Smith, J. O'Grady, Z. Bukhari, C.R. Fricker, J.S. Rosen, and J.L. Clancy. 2000. Inter-laboratory comparison of the CD-1 neonatal mouse dose response model for Cryptosporidium parvum oocysts. *J. Euk. Microbiol*. 47(3): 294-298.

Kruithof, J.C., R.C. van der Leer, and W.A.M. Hijnen. 1992. Practical Experiences with UV Disinfection in the Netherlands. *Aqua*, 41(2):88-94.

LaFrenz, R. 1997. High Intensity Pulsed UV for Drinking Water Treatment. In *Proceedings AWWA Water Quality Technology Conference*. Denver, CO.

Langlaid, B., D.A. Reckhow, D.R. Brink, Eds. 1991, *Ozone in Water Treatment - Application and Engineering*. Denver, CO and Chelsea, MI.: AWWA Research Foundation and Lewis Publishers, Inc

Li, H., G.R Finch, and M. Belosevic. 1999. Effect of Water Temperature on the Sequential Inactivation of *Cryptosporidium* Using Ozone Followed by Free Chlorine. In: *Proceedings AWWA Water Quality Technology Conference*. Tampa, Florida. Oct. 31-Nov. 3, 1999.

Linden, K.G. and J.L. Darby. 1997. Estimating Effective Germicidal Dose from Medium Pressure UV Lamps. ASCE: *Journal of Environmental Engineering*, 123(11) 612-619.

Linden, K. G. and A. A. Mofidi, 1999. Measurement of UV Irradiance: Tools and Considerations. In *Proceedings AWWA Water Quality Technology Conference*, Tampa, FL, Oct 31-Nov 3.

Linden, K.G., G.A. Shin, and M.D. Sobsey. 2000. Relative efficacy of UV wavelengths for the inactivation of *Cryptosporidium parvum*. In *Proceedings of Disinfection 2000, WEF Specialty Conference*.

Linden, K., and J. Christensen, 2001. Ultraviolet Disinfection of Unfiltered Drinking Water: Particle Impacts. *Proceedings of the First IUVA International Congress*. Washington, D.C. June

Linden, K.G., 2001. Progress Reports on the AwwaRF and WERF Collaborative Research Projects on UV Disinfection Techniques for Pathogens. Denver, CO.: AwwaRF and AWWA.

Linden, K., E. Rosenfeldt, S. Johnson, and B. Melcher. 2002. Direct UV and UV Oxidation Processes for Treatment of Taste and Odor Causing Compounds in Water. In Proc. of the AWWA Water Quality Technology Conference. Seattle, Wash.: AWWA.

Lindenauer, K. and J. Darby. 1994. Ultraviolet disinfection of wastewater: Effect of Dose on Subsequent Photoreactivation. *Wat. Res.*, 28(4): 805-817.

Lorenzo-Lorenzo, M.J. 1993. Effect of Ultraviolet Disinfection of Drinking Water on the Viability of Cryptosporidium parvum Oocysts. Journal of Parasitology, 79(1):67-70.

Mackey, E.D, R.S. Cushing, and G.F. Crozes. 2000. UV Disinfection systems for the inactivation of *Cryptosporidium*: Evaluating practical implementation issues, Periodic Report #3, Project 2623, AwwaRF, Denver, Colo.

Maier, A., D. Tougianidou, A. Wiedenmann, K. Botzenhart. 1995. Detection of Poliovirus by Cell Culture and by PCR after UV Disinfection. *Wat. Sci. Tech*., 31 (5-6):141-145.

Malley, J.P., Jr., J.P. Shaw, and J.D. Ropp. 1996. Evaluation of the By-Products Produced by the Treatment of Groundwaters with Ultraviolet Radiation (UV) Radiation and Post Disinfection Following Irradiation. Denver, CO: AWWA and AwwaRF.

Malley, J.P., Jr. 1998. Bench-Scale Collimated Beam Studies to Determine Required UV Dosage for North Kingstown, RI. Technical Report to Fay, Spofford and Thorndike, Inc.

Malley, J.P., Jr., G.L. Snicer, G. Crozes, 1998. Progress Reports on the AwwaRF/EPRI Collaborative Project: Demonstration of Full-Scale UV for Groundwater Disinfection. Denver, CO.: AwwaRF and AWWA.

Malley, J.P. 2000a. *Reduction of Microbial and Chemical Pollution of Coastal Waters Using Pulsed-UV Disinfection*. Interim Report, NOAA.

Malley, J. 2000b. *Demonstration scale UV disinfection study*. Denver, CO.: AwwaRF and AWWA.

Malley, J.P., Jr. 2001. "Engineering of UV Disinfection Systems for Drinking Water," IUVA News, Vol. 3, No.3, p. 7-10, May 2001. IUVA News, Vol.2, No. 3, p. 8-12, June 2000

Malley, J., L.Passantino, Knudson, M., Ward, R., and Kim, J. 2001. The Effects of Turbidity and Algae on the Dose Requirements for Ultraviolet Systems Treating Unfiltered Drinking Water Supplies. In *Proceedings of the First IUVA International Congress.*

Malley, J.P., Jr., B. Towsend, N. Ballester, 2001a. Progress Reports on the NOAA Project: Pulsed UV Disinfection for Reduction of Microbial Pathogen Loads to Coastal Waters and Estuaries, NOAA-CICEET, Durham, NH.

Malley, J.P., Jr., B. Towsend, N. Ballester, 2001b. Progress Reports on the AwwaRF/WERF/USEPA Collaborative Project: Innovative UV Disinfection Techniques for Pathogens. Denver, CO.: AwwaRF and AWWA.

Malley, J.P., Jr. and N.A. Ballester, 2004 "Reduction in Microbial and Chemical Pollution of Coastal Waters Using UV Disinfection", CICEET Final Report, NOAA-CICEET Publication, Durham, NH

Marinas, B.J., J.L. Rennecker, K.M. Ruffelland, and S.A. Rubin. 1999a. Inactivation of *Cryposporidium Parvum* Oocysts with Ozone, Chlorine Dioxide, Free Chlorine, and Monochloramine. In *Proceedings of International Conference on Minimizing the Risk from Cryptosporidium and other Waterborne Particles.* IAWQ/IWSA, *Abstract Proceedings*, Paris, France. April 19-23.

Marinas, B.J., J.L. Rennecker, S. Teefy, and E.W. Rice. 1999b. Assessing Ozone Disinfection with Nonbiological Surrogates. *Journal AWWA*, 91 (9):79-89.

McNair, Scott D.B., E.C. Lesher. 1963. Effect of Ozone on Survival and Permeability of *Escherichia Coli. Journal of Bacteriology*, 85: 567-576.

Meng, Q.S. and C.P. Gerba. 1996. Comparative inactivation of enteric adenovirus, poliovirus and coliphages by ultraviolet irradiation. *Wat. Res.* Vol.30,No. 11pp. 2665-2668

Mofidi, A.A. and Linden, K.G. 1998. Disinfection Using High Intensity Pulsed-Ultraviolet Irradiation. In *Proceedings AWWA Water Quality Technology Conference.* San Diego, Calif., Nov 1-4

Mofidi, A. H. Baribeau, J. Green. 1999. Inactivation of *Cryptosporidium parvum* with polychromatic UV systems. In *Proceedings of the AWWA Water Quality Technology Conference.* Tampa, Fla, Oct 31 – Nov 3

Mofidi, A.A., C.I. Chou and B.M Coffey. 2001. Disinfection of *Giardia lamblia* and *Giardia muris* cysts by UV Light. *First Annual International Ultraviolet Association International Congress.* Washington, D.C. June

Montgomery Watson. 1994. *A Comparative Study of UV and Chlorine Disinfection for Wastewater Reclamation*, Report for Elsinore Valley Municipal Water District and National Water Research Institute.

Morowitz, H. J. (1950). "Absorption Effects in Volume Irradiation of Microorganisms." Science 111: 229.

Motoyama, N., K.Ozawa, D.Chikuma, T.Hirata, K.Takahashi, T.Moniwa, and M.Kaneko. 1999. Inactivation of *Cryptosporidium parvum* Oocysts by Ozone. In *Proceedings of the 14th Ozone World Congress.* London, United Kingdom, International Ozone Assoc.

Meulemans, C.C.E. 1987. *The basic principles of UV-sterilization of water.* In: Ozone Ultraviolet Water Treatment, Aquatec Amsterdam, Paris: International Ozone Association, B.1.1-B.1.13.

National Sanitation Foundation, 1991, *Standard 55: Ultraviolet Microbiological Water Treatment Systems* 1991 NSF Joint Committee on Drinking Water Treatment Units. Ann Arbor, Mich..

Oppenheimer, J.A. 1997. Results of bench-scale testing for Massachusetts Water Resources Authority. Montgomery Watson Draft Technical Memorandum. Nov. 14, 1997.

Oppenheimer, J., M. Aieta, J. Jacangelo, R. Trussell, and I. Najm, 1998. Comprehensive Evaluation of *Cryptosporidium* Inactivation in Natural Waters. In *Proceedings AWWA Water Quality Technology Conference.* San Diego, Cal

Oppenheimer, J., R. Trussell, I Najm, M. Aieta, and J. Jacangelo, 1999. CT Requirements for Ozone Inactivation of *Cryptosporidium*, In *AWWA International Symposium on Cryptosporidium and Emerging Waterborne Pathogens: Extended Abstracts*, Milwaukee, Wis., Aug 29-Sept 1

Oppenheimer, J.A., E.M. Aieta, R.R. Trussell, J.G. Jacangelo, and I.N. Najm. 2000. *Evaluation of Cryptosporidium Inactivation in Natural Waters*. AwwaRF, AWWA. Denver, Colo.

Oppenheimer, J. (2000) Personal Communication.

Owens, J.H., R.J. Miltner, F.W. Schaefer III, and E.W. Rice. 1994. Pilot-Scale Ozone Inactivation of *Cryptosporidium* and *Giardia*. In *Proceedings AWWA Water Quality Technology Conference*, San Francisco, Calif., Nov 6-10

Owens, J.H., R.J. Miltner, T.R. Slifko, and J.B. Rose. 1999. *In vitro* Excystation and Infectivity in Mice and cell Culture to Assess Chlorine Dioxide Inactivation of *Cryptosporidium* Oocysts. In *Proceedings AWWA Water Quality Technology Conference*. Tampa, Florida, Oct 31-Nov 3

Palmateer, G. and E. Whitby. 1987. Ultraviolet disinfection: its Effects on *Escherichia coli* and Bacteriophages as Indicators of Disinfection Efficiency of Wastewater. Presented at *Technology Transfer Conference*, Toronto.

Parker J. & Darby, J.L. 1995. Particle Associated Coliform in Secondary Effluents: Shielding from Ultraviolet Light Disinfection. *Water Environment Research*, 67:7:1065.

Parrotta, M.J. and F. Bekdash. 1998. UV Disinfection of Small Groundwater Supplies. *Jour. AWWA*, 90(2): 71-7

Qualls, R.G. and J.D. Johnson. 1983. Bioassay and Dose Measurement in UV Disinfection. *Applied Environmental Microbiology*, 45:872-877.

Rennecker, J.L., B.J. Marinas, J.H. Owens, and E.W. Rice. 1999. Inactivation of *Cryptosporidium parvum* Oocysts with Ozone. *Water Research*, 33 (11):2481-2488.

Rice, E.W. and J.C. Hoff. 1981. Inactivation of *Giardia lamblia* cysts by Ultraviolet Irradiation. *Jour. Appl. and Envir. Microbiology*, 42(3) 546-547.

Robertson, L.J., A.T. Campbell, and H.V Smith. 1993. In vitro excystation of *Cryptosporidium parvum*. *Parasitology* 106:13-19.

Rubin, L.B., 1981. The Effects of High Intensity UV Radiation on Nucleic Acids and Their Components – I. Thymine. *Photochem. Photobiol.*, 34:339-345.

Ruffell, K.M., J.L. Rennecker, and B.J. Marinas. 1998. Inactivation Kinetics of *Cryptosporidium parvum* Oocysts with Chlorine Dioxide. In *Proc. 1998 AWWA Water Quality Technology Conference*. San Diego, Calif.

Sauch J.F., D. Flanigan, M.L. Galvin, D. Berman,W. Jakubowski, 1991 "Propidium iodide as an indicator of Giardia cyst viability". *J. Appl Environ Microbiol* 57:3243-7.

Scheible, O.K., M. Casey, and A. Forndran. 1986. *Ultraviolet Disinfection of Wastewaters from Secondary Effluent and Combined Sewer Overflows*. EPA-600/2-86/005. PB86-145182, U.S. Environmental Protection Agency, Cincinnati, Ohio.

Shaw, J.P. 1994. The Effects of Ultraviolet (UV) Irradiation on Organic Matter in Natural Waters. Masters Thesis. Department of Civil Engineering, University of New Hampshire, Durham, NH.

Shin, G., K.G. Linden, and M.D. Sobsey. 1999. Inactivation of *Cryptosporidium* oocysts with Low Pressure UV Radiation. In *Proceedings of the Water Quality Technology Conference*. Tampa, Fla., Oct 31 – Nov 4

Shin, G.A. K.G.Linden, and M.D. Sobsey. 2000a. Comparative Inactivation of *Cryptosporidium parvum* oocysts and Coliphage MS-2 by Monochromatic UV Radiation. In *Proceedings of the Disinfection 2000 WEF Specialty Conference.*

Shin G., K. G. Linden, G. Faubert, and M. D. Sobsey. 2000b. Inactivation of *Cryptosporidium parvum* Oocysts and *Giardia lamblia* Cysts by Monochromatic UV Radiation. In *Proc. IWA First International Congress, Health Related Water Microbiology.*, Paris, France, July 3-7

Shin G., K. G. Linden, M. J. Arrowood, G. Faubert and M. D. Sobsey. 2001a. DNA repair of UV-irradiated *Cryptosporidium parvum* oocysts and *Giardia lamblia* cysts. *First Annual International Ultraviolet Association International Congress.*

Shin, G-A., K.G. Linden, M.J. Arrowood, and M.D. Sobsey. 2001b. Low pressure UV inactivation and DNA repair potential of *Cryptosporidium parvum* oocysts. *Appl. Environ. Microbiol.* 67(7):3029-3032.

Shin G., Personal Communication, Duke University (2001).

Slifko, T.R., D.E. Friedman, J.B. Rose, S.J. Upton, and W. Jakubowski. 1996. Unique Cultural Methods Used to Detect Viable *Cryptosporidium parvum* Oocysts in Environmental Samples. *Water Science & Technology*, 35(11-12):363-368.

Slifko, T.R. D. Friedman, J.B. Rose, and W. Jakubowski, 1997, An In-vitro Method for Detecting Infectious Cryptosporidium Oocysts with Cell Culture, *Appl. Environ. Microbiol.* 63 (9): 3669-3675.

Slifko, T.R., Rose, J.B., Friedman, D.E. 1997. Cryptosporidium Viability and Infectivity for the Water Industry. In *Proceeding of the American Water Works Association Annual Conference*. Atlanta, GA, June 15-19: AWWA.

Slifko, T.R., D.E. Friedman-Huffman, and J.B. Rose. 1998. Comparison of 4 *Cryptosporidium parvum* Viability Assays: DAPI/PI, Excystation, Cell Culture, and Animal Infectivity. In *Proc. AWWA Water Quality Technology Conference*. San Diego, Calif., Nov 1-4

Snicer, G.A., J.P. Malley, A.B. Margolin, and S.P. Hogan. 1998. *Evaluation of Ultraviolet Technology in Drinking Water Treatment*. Denver, Colo.: AwwaRF and AWWA.

Sommer, R., T. Haider, A. Cabaj, W. Pribil and M. Lhotsky (1998) Time dose reciprocity in UV disinfection of water. Poster Presented at the IAWQ Vancouver, Canada.

Soroushian, F. (2000) Personal Communication.

Stewart, M.H., ?? 1993. Microbial and Chemical Implications of Using Ultraviolet Irradiation for Treatment of Biological Filter Effluent. In *Proceedings of the AWWA Water Quality Technology Conference,* Miami, Fla., Nov 14-18

Stewart, M.H. and B.H. Olson.1996. Bacterial resistance to potable water disinfectants, pp. 140-192. In *Modeling Disease Transmission and Its Prevention by Disinfection*. C.J. Hurst (ed.). Cambridge University Press, Cambridge

Stibbs HH, Samadpour M, Manning JF. (1988) Enzyme immunoassay for detection of Giardia lamblia cyst antigens in formalin-fixed and unfixed human stool. *J Clin Microbiol* 26:1665-9.

Suidan, M.T. and B.F. Severin. 1986. Light Intensity Models for Annular UV Disinfection Reactors. *Jour. A.I.Ch.E.*, 32: 1902-1909.

Tosa, K. and T. Hirata. 1998. Photoreactivation of *Salmonella* following UV disinfection. *IAWQ 19th Biennial International Conference - Vol. 10 Health-Related Water Microbiology.*, Vancouver, BC Canada.

Tree, J.A., M.R. Adams, D.N. Lees. 1997. Virus Inactivation During Disinfection of Wastewater by Chlorination and UV Irradiation and the Efficacy of F+ Bacteriophage as a 'Viral Indicator'. *Wat. Sci. Tech.*, 35(11-12)227-232.

U.S. Environmental Protection Agency, Office of Research and Development, 1986. *Design Manual: Municipal Wastewater Disinfection*. EPA 625/1-86/021. Cincinnati, OH: USEPA.

U.S. Environmental Protection Agency, 1989, 40 CFR Parts 141 and 142, National Primary Drinking Water Regulations: Filtration and Disinfection: Turbidity, *Giardia lamblia*, Viruses, *Legionella*, and Heterotrophic Bacteria. Final Rule. *Federal Register*, 54:124: 27486, June 29, Washington, D.C.

U.S. Environmental Protection Agency. 1991. *Guidance Manual for Compliance with the Filtration and Disinfection Requirements for Public Water Systems Using Surface Water Sources* Contract No 68-01-6989, March, Washington, D.C.

U.S. Environmental Protection Agency. Office of Groundwater and Drinking Water. 1997. DRAFT Alternative Disinfectant and Oxidants Guidance Manual. (CN:68-C6-0059) Washington, D.C.

U.S. Environmental Protection Agency, 1998, 40 CFR Parts 9, 141, and 142, National Primary Drinking Water Regulations: Disinfectants and Disinfection Byproducts; Final Rule, *Federal Register* 63:241: 69390-69476, December 16, Washington, D.C.

U.S. Environmental Protection Agency, 1998, 40 CFR Parts 9, 141, and 142, National Primary Drinking Water Regulations: Interim Enhanced Surface Water Treatment; Final Rule, *Federal Register*, 63:241: 69478-69521, December 16, Washington, D.C.

U.S. Environmental Protection Agency/National Sanitation Foundation. 1998. Environmental Technology Verification Program: Draft Report: Performance Report of the Sentinel™ Ultraviolet Tower for Inactivation of Giardia and Cryptosporidium.

U.S. Environmental Protection Agency, 2001, LT2ESWTR Preproposal Draft Regulatory Language for Stakeholder Review, November 27, Washington, D.C.

U.S. Environmental Protection Agency. 2003a. 40 CFR Parts 141 and 142. National Primary Drinking Water Regulations: Long Term 2 Enhanced Surface Water Treatment Rule; Proposed Rule. *Federal Register* (Part II), 68(154):47709-47711, August 11, Washington, D.C.

U.S. Environmental Protection Agency, Office of Water, 2003b. *Ultraviolet Disinfection Guidance Manual - Draft.* EPA 815-D-03-007, June, Washington DC

U.S. Environmental Protection Agency, Office of Water, 2003c, *Long Term 2 Enhanced Surface Water Treatment Rule Toolbox Guidance Manual - Draft*, EPA 815-D-03-009, June, Washington DC

Venkatesan, N., and Reckhow, D.A., 2003a, UMass *Progress Report #1* Disinfectant Bench Testing and DBP Analytical Support, February, Amherst, Mass.

Venkatesan, N., and Reckhow, D.A., 2003b, UMass *Progress Report #2* Disinfectant Bench Testing and DBP Analytical Support, March, Amherst, Mass.

Venkatesan, N., and Reckhow, D.A., 2003c,UMass *Progress Report #3* Disinfectant Bench Testing and DBP Analytical Support, May, Amherst, Mass.

Venkatesan, N., and Reckhow, D.A., 2003d, UMass *Progress Report #4* Disinfectant Bench Testing and DBP Analytical Support, May, Amherst, Mass.

Venkatesan, N., and Reckhow, D.A., 2003e, UMass *Progress Report #5* Disinfectant Bench Testing and DBP Analytical Support, June, Amherst, Mass.

Venkatesan, N., and Reckhow, D.A., 2003f, UMass *Progress Report #6* Disinfectant Bench Testing and DBP Analytical Support, July, Amherst, Mass.
Venkatesan, N., and Reckhow, D.A., 2003g, UMass *Progress Report #7* Disinfectant Bench Testing and DBP Analytical Support, August, Amherst, Mass.
Venkatesan, N., and Reckhow, D.A., 2003h, UMass *Progress Report #8* Disinfectant Bench Testing and DBP Analytical Support, September, Amherst, Mass.
Venkatesan, N., and Reckhow, D.A., 2003i, UMass *Progress Report #9* Disinfectant Bench Testing and DBP Analytical Support, November, Amherst, Mass.
Venkatesan, N., and Reckhow, D.A., 2003j, UMass *Progress Report #10* Disinfectant Bench Testing and DBP Analytical Support, November, Amherst, Mass.
Venkatesan, N., and Reckhow, D.A., 2003k, UMass *Progress Report #11* Disinfectant Bench Testing and DBP Analytical Support, November, Amherst, Mass.
Venkatesan, N., and Reckhow, D.A., 2003l, UMass *Progress Report #12* Disinfectant Bench Testing and DBP Analytical Support, December, Amherst, Mass.
von Sonntag, C. 1986. Disinfection by Free Radicals and UV Radiation. *Jour. Water Supply*. 4:11-19.
von Sonntag, C. and H.P. Schuchmann. 1992. UV Disinfection of Drinking Water and By-product Formation--Some Basic Considerations. *Aqua*. 41(2):67-74.
Weinberg, H.S., W.H. Glaze, S.W. Krasner, and M.J. Sclimenti, 1993, Formation and Removal of Aldehydes in Plants that Use Ozone, *JAWWA*, Vol. 85, No 5
Whitby, G.E., G. Palmateer, W.G. Cook, J. Maarschalkerweerd, D. Huber, and K. Flood. 1984. Ultraviolet disinfection of secondary effluent. *J. Wat. Pollut. Control Fed,*. 56:844-850.
Wiedenmann, A., B. Fischer, U. Straub, C.-H. Wang, B. Flehmig and D. Schoenen. 1993. Disinfection of Hepatitis A virus and MS-2 coliphage in water by ultraviolet irradiation: Comparison of UV-susceptibility. *Wat. Sci. Tech*., 27(3-4):335-338.
Wilson, B.R., P.F. Roessler, E. Van Dellen, M. Abbaszadegan and C.P. Gerba. 1992. Coliphage MS-2 as a UV water disinfection efficacy test surrogate for bacterial and viral pathogens. In the *Proceedings of the Water Quality Technology Conference*. Toronto, Ont.: AWWA. Nov 15-19
Witherell, L.E., R.L. Solomon, and K.M. Stone. 1979. *Ozone and Ultraviolet Radiation for Small Community Water Systems*, EPA 600/2-79-060. Washington, D.C.: USEPA.
Yanko, B. (2000) Personal Communication.
Yip, R.W. and D.E. Konasewich. 1972. Ultraviolet Sterilization of Water – Its Potential and Limitations. *Water Supply and Pollution Control (Canada)*. 14:14-18.
Zemke, V., L. Podgorsek and D. Schoenen (1990) Ultraviolet disinfection of drinking water 1. Communication: Inactivation of *E. coli* and coliform bacteria. *Zbl. Hyg*. Vol. 190, pp. 51-61

ABBREVIATIONS

A	absorbance
AOC	assimilable organic carbon
ATCC	American Type Culture Collection
AwwaRF	Awwa Research Foundation
°C	degrees Celsius
CEC	Clancy Environmental Consultants
cm	centimeters
CFU/mL	colony forming units per milliliter
CT	concentration over time
DAPI	4′,6-Diamino-2-phenylindole
DI	de-ionized
DNA	deoxyribonucleic acid
DBP	disinfection by-product
E. coli	*Escherichia coli*
gpm	gallons per minute
HAA	haloacetic acid
HPC	heterotrophic plate count
Hz	hertz
I	irradiance intensity
IESWTR	Interim Enhanced Surface Water Treatment Rule
IT	irradiance over time
log	logarithm, base 10
LPHO UV	low pressure, high output ultraviolet light
LP-UV	low pressure ultraviolet light
LT1ESWTR	Long Term 1 Enhanced Surface Water Treatment Rule
LT2ESWTR	Long Term 2 Enhanced Surface Water Treatment Rule
µL	microliters
µW/cm^2	microWatts per square centimeter
MWRA	Massachusetts Water Resources Authority
mA	milli-Amp
mg/L	milligrams per liter
mg·min/L	milligrams per liter times minutes
mL	milliliter
mJ/cm^2	milliJoules per square centimeter
mW/cm^2	milliWatts per square centimeter

MPSS	Multiple Point Source Summation Approximation
MP-UV	medium pressure ultraviolet light
MS2	RNA virus that infects *E. coli*
nm	nanometers
NIST	National Institute for Standards and Technology
NOM	natural organic matter
NSF	National Sanitation Foundation, International
%	percent
PI	propidium iodide
Q	flow
RED	reduction equivalent dose
RO	reverse osmosis
RNA	ribonucleic acid
scfh	standard cubic feet per hour
SWTR	Surface Water Treatment Rule
T	transmittance
T_{10}	detention time for 90 percent of the flow in a basin
T_{theory}	theoretical detention time within a basin, volume divided by flow
TFTC	too few to count
THM	trihalomethane
TNTC	too numerous to count
TOX	total organic halides
UNH	University of New Hampshire
UNH-ERG	Inuversity of New Hampshire Environmental Research Group
USEPA	United States Environmental Protection Agency
UV	ultraviolet light
UV_{254}	ultraviolet light absorbance at the wavelength 254 nanometers
UVDGM	Ultraviolet Disinfection Guidance Manual
UVT	ultraviolet light transmittance
VIS	visible wavelengths
V	volume
Θ_H	hydraulic detention time

IWA Publishing's authorised EU representative for General Product Safety Regulations is Diane D'Arras, 15 rue Duret, 75116 Paris, France, e-mail: safety@iwap.co.uk.

Printed and bound by CPI Group (UK) Ltd, Croydon, CR0 4YY
17/09/2026
02218811-0014